ENCYCLOPAEDIA OF GENETICS - IV

ORIGIN AND EVOLUTION OF GENETICS

By

Dr. Arvind N. Shukla

School of Studies of Zoology & Biotechnology
Vikram University
Ujjain (M.P.)
(India)

DISCOVERY PUBLISHING HOUSE PVT. LTD.
NEW DELHI-110 002

First Published - 2009

Reprinted - 2016

ISBN: 978-81-8356-395-6

Published by:

DISCOVERY PUBLISHING HOUSE PVT. LTD.

4383/4B, Ansari Road, Darya Ganj
New Delhi-110 002 (India)
Phone: +91-11-23279245, 43596064-65
Fax: +91-11-23253475
E-mail: discoverypublishinghouse@gmail.com
sales@discoverypublishinggroup.com
web: www.discoverypublishinggroup.com

Printed at:
Infinity Imaging Systems
Delhi

Preface

The present title "*Origin and Evolution of Genetics*" is an exciting, and dynamic branch of science and offers the finest approach to teaching genetics through the integration of the molecular and chemical subdisciplines. It prepares the students to learn to formulate genetic hypothesis and apply critical thinking skill necessary for problem solving, while also gaining a sense of the social and historical context in which genetics has developed. This text also has a completely novel way to illustrate the one or two experiments in each chapter that are rigorously examined according to the scientific method. It starts with the premise that the syllabus for a university course in genetics should reflect the major research issues of the new millennium rather than those topics that were in vogue during the last decades of nineteenth century.

The text covers both the basic and practical aspects of Genetics. Fundamental knowledge is developed within the context of applied relevance. Principles are supplemented with examples. This is done to maintain student interest, which is essential for learning any subject.

In the preparation of this book large number of books and research papers have been consulted. So no authenticity is claimed.

The author expresses his gratitude to Mr. Wasan and staff of M/s Discovery Publishing House Pvt. Ltd. for their whole hearted co-operation in the publication of this book.

The author tried hard to be accurate and upto date in statement and realises the impossibility of completely avoiding errors therefore, the author will greatly appreciate having his attention called to any questionable statement.

Author

CONTENTS

1

INTRODUCTION

In July 1996 a nearly complete human skeleton was found by two college students on the bank of the Columbia River near the town of Kennewick, Washington. Human remains so encountered are always of concern—"Who was it?" and "Who did it?"—so the students called the police. In an effort to answer those two questions the police turned the bones over to the local coroner. Burial grounds of Native Americans are sometimes encountered in that part of the Northwest, and the Native American Graves Protection and Reparation Act, passed in 1990, required that any such remains be returned for burial to the tribe to which they belonged. The bones appeared to be of great age, and so it was assumed that the skeleton must be one of a Native American, since settlers of European origin reached the West Coast only a few centuries ago. But closer study suggested otherwise.

To solve the puzzle, the bones were examined by an anthropologist, James Chatters, a specialist in skeletal remains of human beings. Such professionals can determine sex, size, age, cause of death, and racial type with considerable accuracy. Examination showed the skeleton to be that of a 50-year-old male of medium build, his teeth well worn and a stone arrowhead imbedded in his hip bone. Radiometric methods determined that the man died about nine thousand years ago—long before human beings of European origin first arrived in the New World, according to conventional historical accounts. Yet the skeleton had Caucasoid features. Chatters took the bones to another anthropologist for an opinion, without giving a hint of his analysis, and was told that the skeleton was of a *Caucasian male*. Even when Chatters revealed the age of the bones, the second anthropologist stuck to her original identification. A third anthropologist familiar with the skeletal features of modern tribes of Native Americans concluded that the skeleton could not be assigned to any one of them.

Finding the nearly nine-thousand-year-old skeleton of a Caucasoid male in any part of the New World is puzzling in the extreme. Traditionally, anthropologists have thought that the first human beings to inhabit the Western Hemisphere crossed from Siberia to Alaska about 15,000 years ago. Some now believe that the event occurred much earlier, but in any case these immigrants from Siberia were of the Mongolian racial type, as are Native Americans—not Caucasoids. Anthropologists know that a few nameless fishermen from Western Europe came to the eastern shores of the New World before Columbus's arrival in 1492, as did the Vikings, but Caucasians did not come in large numbers until early in the sixteenth century.

So who was the Caucasoid Kennewick Man who arrived thousands of years before the Vikings, Columbus, Cortez, and Pizarro? Needless to say, this is a most exciting and important question, not only for anthropologists and historians but for many nonprofessionals as well. There have even been some speculations that Caucasoid people may have been the original inhabitants of the New World. Thus, for scientists and others interested in such historical questions, further studies on Kennewick Man are overwhelmingly important.

For a while it looked as though these investigations would never take place. In an effort to comply with the federal Native American Graves Protection and Reparation Act, the Army Corps of Engineers assumed control of the bones, placed them in a vault, and refused to allow any further examination of them by scientists. The Umatilla tribe, who live near the site of discovery, asked to have the bones returned to them, in which case the skeleton would be secretly buried and never be available for study. A group of anthropologists went to court to stop the Corps from complying with the tribe's request. The anthropologists claimed that the Umatillas were not in that part of the Northwest when Kennewick Man lived; hence, he could not be one of their ancestors. And of course his Caucasoid skeletal features led to the same conclusion. Thus, the available scientific evidence is that Kennewick Man was not a Umatillan or any other Native American.

In response to that hypothesis, a leader of the tribe, Armand Minthorn, stated this position: Our elders have taught us that once a body goes into the ground, it is meant to stay there until the end of time. . . . If this individual is truly over 9,000 years old, that only substantiates our belief that he is Native American. From our oral histories, we know that our people have been part of this land since the beginning of time. We do not believe that our people migrated here from another continent, as the scientists do. . . . Scientists believe that because the individual's head measurements do not match ours, he is not Native American. Our elders have told us that Indian people did not always look the way we look today. Some scientists say that if this individual is not studied further, we, as Indians, will be destroying evidence of our history. We already know our history. It is passed on to us through our elders and through our religious practices.

As of early 2001, the matter remains unsettled. Nevertheless the two perspectives—the anthropologists' and the Native Americans'—provide a classic example of two polar points of view that will be analyzed throughout this book. One point of view rests on the questions and methods of science. The other rests on cultural beliefs that have been passed down from generation to generation. One side seeks a solution to a problem of intense interest to scientists and historians; the other side does not recognize that there is a problem that needs a solution.

This dispute over the future of Kennewick Man represents a clash between two immiscible patterns of thought. Another example comes from Afghanistan, where the Taliban, a Muslim sect, have gained control of much of the nation and are enforcing conformity to the Sharia, sacred Islamic law. According to this code, thieves must be punished by having their hands and feet cut off; couples caught in adulterous acts must be stoned to death; and if women do not cover themselves from head to foot, the young Taliban enforcers deal them a severe fogging. The young zealots have even beaten women for wearing white socks or plastic sandals.

The head of the General Department for the Preservation of Virtue and Prevention of Vice, Alhaj Maulavi Qalamuddin, explains the Taliban point of view: "Some women want to show their feet and ankles. They are *immoral women*. They want to give a hint to the opposite sex." This must be controlled to "prevent impure thoughts in men"; "if we consider sex to be as dangerous as a loaded Kalashnikov rifle, it is because it is the source of all immorality." The rules of the Sharia relating to women are harsh in other respects, by late-twentieth-century standards in the West. Women are prohibited from working or obtaining an education or even receiving medical treatment, and after puberty they are almost entirely secluded in their homes.

Such behaviour toward women is not acceptable in most nations today, including most nations where Islam is the predominant religion. In this type of conflict modern values concerning human rights, gender equality, and civil liberties clash with religious doctrines that have been handed down for thousands of years. Such doctrines are accepted as "*true*" by the culture that inherits them and are highly resistant to change. In most instances, there is no way to adjudicate a conflict between these two systems of belief with evidence acceptable to each side.

Many—probably most—of the problems between nations, as well as the problems among the people of a single nation, stem from taking different points of view toward the same problem. Anthropologists using the methods and data of science propose one course of action for dealing with Kennewick Man, whereas the Native Americans relying on their received traditions propose another. Granted, only the scientists offer the possibility of reaching a decision about the identity of the remains based on confirmable data rather than on faith, but pursuing this course requires that both sides accept the validity of the anthropologists' assumptions, methods, and evidence; such acceptance by the *Umatilla tribe* seems unlikely. Consequently, in keeping with American jurisprudence, a federal judge will decide which point of view will prevail. By contrast, science has nothing useful to say when—as in the case of the Afghan women—the clash is between human rights and divine law. That conflict is between two different belief systems, each accepted as a matter of faith or principle, with one being far more restrictive on the lives of women than the other.

Creationism versus Evolution

The immiscible patterns of thought that concern us in this book are those of creationists and evolutionists. Christian fundamentalists accept without question that divine creation is the explanation for the diversity of life we see today—the many different species of plants, animals, fungi, and microorganisms that flourish around the globe. Their position is based on their reading of Genesis, with its familiar story of the creation week—six days during which God created all of nature. On the first day God created heaven and earth and light and darkness; on the second He made the firmament and divided the waters; on the third day He separated land from the seas and created the land plants; on the fourth day He created the sun, moon, and other celestial bodies; on the fifth day animals in the sea and birds in the air came into being; on the sixth day the land animals appeared, and God also created, in his own image, two human beings. Until recently the accepted date for creation, based on a tally of the generations ("begats") listed in the King James version of the Bible, was 4004 b.c. Creationists assume that all creatures living today are the same as when they were created. That is, there has been no evolution.

The scientific version of these events is quite different. Tentative estimates place the origin of the universe in the neighbourhood of 15 billion years ago. The *sun*—a second-generation star—and its orbiting planets formed about 4.5 billion years ago from the interstellar debris created by the explosion of massive first-generation stars. The Earth started out in a molten state but cooled enough to form a solid crust by about 3.8 billion years ago.

The first evidence of life dates to about 3.5 billion years ago and consists of very simple cells without a nucleus (*prokaryotes*), much like *bacteria* living today. The oldest known cells with nuclei (eukaryotes) date to about 2.7 billion years ago. The earliest multicelled organisms discovered so far date to about 1.7 billion years ago. *Fossils* of the first members of the animal kingdom date to about 650 million years ago. At the beginning of the Cambrian period, about 570 million years ago, animal life became abundant and highly diversified; the fossil record from this period is much better known than that of earlier times. From that point until today, there has been an incredible evolution of life (both plants and animals), with different species appearing, flourishing, and then becoming extinct or evolving into still other species.

The *evolutionists*' and the *creationists*' accounts for the origins and diversity of life could hardly be more incompatible. Strict creationists base their account on faith—a belief that Genesis was divinely inspired and provides the only true explanation for the origins of the universe, living creatures, and the many variations in organisms that we observe today and find in the fossil record. The creationists' pattern of thought begins with the answer and then seeks to explain the world in terms of that answer. Scientists work in the opposite direction. They study the universe and its earthly inhabitants and on the basis of observations and experiments propose a rational account for the past and the present.

Scientists as a group, including those who adhere to religious beliefs, do not understand why any person familiar with the data would reject an explanation based on confirmable knowledge and accept instead a supernatural concept based on faith alone. Similarly, deeply committed fundamentalists wonder why anyone would reject the Word of God in favour of what a bunch of scientists has to say.

The long strands of human history have seen many conflicts between science and religion, and sadly, they have often turned violent and bloody. It is fascinating to consider how individuals come to hold such conflicting explanations for the same phenomena and why they hold them so tenaciously. We know that parents and society are remarkably efficient in transmitting patterns of thought and behaviour to successive generations. This cultural inheritance sets up rules for behaviour that help individuals get along within their group.

It also provides each generation with an avenue for learning new things, and it creates order within the society. But beyond these practical advantages, a culture's unique belief systems may retain their enormous power from generation to generation in large part because they supply impressionable young people with answers to many questions they quite naturally ask, such as "Where did I come from?" "Who made me?" "Who will take care of me?" "Why do people die?" "What happens to me after I die?"

Such inquisitiveness seems to be part of our human inheritance. For hundreds of thousands of years, early human populations were illiterate and encapsulated in small tribes whose very survival was regularly endangered. Human beings lived a marginal existence like all other animals, dependent on the ability of the environment to sustain them and on their skills and knowledge of how to obtain food, water, and shelter. But within every society, some individuals must have exhibited a much stronger desire than their peers not just to survive from moment to moment but to understand themselves and the world around them. They asked questions and sought answers.

According to anthropologists who have studied hunter-gatherer cultures, the explanations inquisitive tribe members come up with usually include both natural and supernatural elements. Natural things and processes are those that can be observed: wind, rain, birth, death, animals, plants, fire, night, day, and the seasons. But for each of these observable entities, a supernatural element of some kind usually figures in the explanation as well. For example, although death is now accepted by most people in Western societies as due to natural causes such as disease, accidents, or the ravages of age, people in some parts of the world still believe that death results from the displeasure of a god or spirit or the effects of a curse such as the evil eye.

The tenacity with which people hold onto beliefs in the supernatural or paranormal has been the subject of much scientific investigation. The results are complex and unexpected. Two psychologists, Barry Singer and Victor A. Benassi (1981, 49–51), made an extensive study of occult beliefs in the United States and offer this summary: Far from being a "*fad*," preoccupation with the occult now forms a pervasive part of our culture. *Garden-variety* occultisms such as astrology and ESP [extrasensory perception] have swelled to historically unprecedented levels. . . . Belief in ESP, for instance, is consistently found to be moderate or strong in 80–90% of our population; . . . in one survey it ranked as our most popular supernatural belief, edging out belief in God in strength and prevalence.

Experiments which have attempted to encourage disconfirmation of occult or illusory beliefs by motivating subjects to think through their judgments more carefully . . . have uniformly revealed an astonishing resistance to change of such beliefs. . . . [Such] stubbornness of illusory and occult beliefs is typical rather than exceptional.

By way of explaining the origin of beliefs in the paranormal, Singer and Benassi point out that if human beings do not understand the reason for a given event, they tend to invent one, and the kind of reason they invent will depend on the intellectual baggage they carry with them. A simple, rational, natural explanation might be preferred, but if that cannot be developed, the need to explain remains and supernatural causes may be invoked. Thus, according to these investigators, one factor in the great increase of interest in the occult in the United States is inadequate science education in the schools. Over half of the students they tested did not know, for example, that the level of water in a partially filled glass remains parallel to the Earth's surface as the glass is tipped.

Singer and Benassi suggest that the "*iffiness*" of science is a second factor that weighs against the acceptance of scientific explanations over supernatural ones. Scientists tend to resist claiming that their statements are true by any absolute measure, stating only that they represent the best available explanation based on existing data. As Thomson (1997, 219) expresses it, the facts of science are "temporary way-stations on the long path to the refinement of knowledge." For many people, this tentativeness is a pale substitute for the finality and certainty of supernatural explanations.

A third factor that has increased interest in the occult is the media, which report stories about UFOs (*unidentified flying objects*), psychic healing, people who claim that the dead speak through them, and ghosts. Splendid examples of articles of this kind can usually be found in publications stacked at the checkout counters of supermarkets. The media rarely provide scientific analyses of these reports. Added to this "news from the other side" are highly entertaining motion pictures and television dramas that contain supernatural and superhuman feats. And of course even our own dreams can be quite extraordinary and entertaining, carrying us into compelling worlds where the constraints of waking reality no longer apply.

But Singer and Benassi found that the biggest factor by far in people's acceptance of the occult is organized religion. Most people profess a belief in some religion, and essentially all religions accept miracles as a given. This institutionalized belief in miracles, according to Singer and Benassi, has a spillover effect into other realms of life and accounts in large part for people's acceptance of paranormal events ranging from visits by angels to alien abductions.

Just how rigid occult beliefs can be is documented in an interesting experiment done with students in a psychology course at Concordia University in Montreal (Gray 1984). The course was called "The Science and Pseudoscience of Paranormal Phenomena." Students were given a test at the beginning of the class to see whether they believed in ESP, ghosts, or miracles. At the end of the semester they were asked the same questions to see what effect studying these phenomena would have on their beliefs. Then, to get an estimate of the stability of any change in beliefs, the students were asked the same questions a year later. Occult beliefs declined by the end of the course but usually by trivial percentages. After one year, belief in the paranormal had moved back to levels close to those before the course was taken. In some cases it was slightly higher. This one-semester course designed to reduce students' belief in the paranormal was not a success. Lawson and Weser (1990, 589) report similar findings in another study and note that "the less skilled reasoners were more likely to initially hold the nonscientific beliefs and were less likely to change those beliefs during instruction. It was also discovered that less skilled reasoners were less likely to be strongly committed to the scientific beliefs."

Should we conclude that a willingness to accept supernatural explanations over scientific ones has been hardwired into the human brain—that it is "*human nature*" to seek meaning beyond the confines of the natural world? Perhaps, but an alternative—and to my mind more probable—hypothesis is that one's patterns of thought and belief are the result of influences very early in life associated with family, church, friends, community, books, other media, and the schools. Children almost always develop the habits and beliefs of the family and culture to which they belong; indeed, the fidelity of cultural inheritance often seems as strong as the fidelity of genetic inheritance. Most children in the United States grow up in communities where they see many churches but encounter few or no institutions that extol the methods of science. Many television channels they watch feature evangelists expounding on miracles, heaven, and eternal damnation but show few scientists explaining science as a way of knowing. Most people are exposed to stories of creation from childhood, but few of them ever hear scientists' account of evolution.

Natural versus Supernatural Explanations

Natural and supernatural explanations for the way the world works have ebbed and flowed throughout human history, and both are still with us today. It is tempting to imagine a slow progress through time toward rational thought and naturalistic explanations and away from superstition, cults, and magic, but this is not the case. Shortly after *World War II*, the remarkable advance of science and technology elevated respect for the procedures and discoveries of science. But by the late 1960s, many people had become disenchanted with scientific perspectives and skeptical of the ability of scientists to do more good than harm.

Conversely, attempts to exclude the supernatural from explanations of natural events have been an important part of Western civilization from its beginning. As far back as the sixth century b.c., a few Greek philosophers who lived in Ionia on the coast of what is now western Turkey attempted to understand natural processes in terms of natural causes, a firmly rooted characteristic of science to this day. Only fragments remain of what they wrote, and most of the information we have about them comes from the surviving manuscripts of Greek authors who lived several centuries later. But from what scholars can tell, those Ionian thinkers were the earliest scientifically minded philosophers in the West.

The philosophy and science of the classical world of Greece and later of Rome continued this tradition. Aristotle in the fourth century b.c. codified the science of his time; and although later observations showed him to be incorrect in many instances, he sought natural explanations for natural phenomena. Throughout the centuries when Rome ruled the *Mediterranean world*, Western civilization saw a slow improvement in living conditions, technology, and general knowledge, all based to a great extent on a scientific approach to problem-solving. Those trends began to reverse themselves, however, in the fourth century a.d., with the passing of the great Roman emperors. By 476, after a long decline, the Roman Empire in the West fell and its political center moved east to Constantinople (modern Istanbul). The emperor Constantine was a Christian, and the Christian Church became increasingly powerful throughout the Middle East and the Mediterranean. The West had entered the Age of Faith. During the millennium from the fall of Rome to the fall of Constantinople in 1453, intellectual activity centered on the Church and was devoted to Christian practice and doctrine. Individuals who might have been the scientists, political leaders, artists, teachers, and philosophers in a more secular society were almost all occupied in the service of the Church. The literacy and education of the time were centered in the monasteries. Science and technology progressed little, and much of what had been generally known in classical times was forgotten.

The story was quite different farther east, in China, India, and Arabia, where learning flourished during this period. Islamic scholars made great contributions to mathematics and astronomy—a legacy still with us in the Arabic names of the major stars and in Arabic numerals. Eastern medicine was also superior to that in the West. Arabic scholars were greatly interested in what Aristotle, Plato, Galen, and other Greeks had written, and they translated the Greek manuscripts available to them into their own language. Later these works were translated from *Arabic to Latin* and so became available to scholars in the West. The quality of these classical works was so superior to any secular writings of the time that they eventually joined the *Bible* as the basic texts of the *medieval period*.

As writings of Greek and Roman scientists became available, some Christian scholars saw the need to adjust them to Christian beliefs. That great doctor of the Church Saint Thomas Aquinas (1225–1274) tried, for example, to adjudicate between the Averroists, who, following Aristotle, held that faith and reason were two absolutely distinct ways of thinking, and the Augustinians, who, following Saint Augustine (354–430 a.d.), believed that faith always precedes reason. Aquinas suggested a compromise: the truths of faith complement the truths of reason. This puzzling solution had advantages: in the short term, it protected the *Averroists* from being classed as *heretics*, with possibly grisly consequences; and in the long term, it saved Aristotle's work from banishment and kept the cause of reason alive in the Age of Faith.

By around 1400 a.d., another change in the intellectual climate began to be felt, leading to the period historians have traditionally called the Renaissance. This "*rebirth*" started in Italy in the late fourteenth century and then spread throughout Western Europe. Slowly but inexorably, humanistic themes replaced religious subjects in literature and the arts. Henry the VIII in England and Martin Luther and John Calvin on the Continent lessened the iron grip that the Catholic Church had exercised over those seeking new knowledge. Science and secular philosophy became acceptable pursuits once again. The year 1543 saw the greatest breakthrough in the cause of science, as three important intellectual works were published: the writings of the Greek physicist and mathematician of the third century b.c., Archimedes, which provided a basis for the later work of Galileo and Newton; Andreas Vesalius's text on human anatomy; and De Revolutionibus Orbium Coelestrium by Nicolaus Copernicus, a Polish astronomer and clergyman.

Both Vesalius and Copernicus proposed ideas that flew in the face of beliefs the Catholic Church had upheld almost without question for a millennium. These beliefs were based on the writings of Galen on human anatomy and of Ptolemy on astronomy. When Michael Servetus, a Spanish physician, theologian, and scholar, made observations suggesting that Galen was not always correct in his anatomical descriptions, the Church had Servetus arrested, investigated, and in 1553, burned at the stake with a copy of one of his offending publications hung around his neck. When Giordano Bruno, a Dominican monk, rejected the Church-approved system of Ptolemy in which the heavens circled the Earth and accepted the system of Copernicus in which the Earth circled the sun, he too was burned alive, in 1600. Being burned alive is one of the most painful ways to die and, hence, appropriate for those who challenged authority in those times.

Still, a revolution in science had burst on the scene with a vigor the Church could not stamp out. By the time the seventeenth century came to a close, William Harvey (1578–1657) had described the circulation of the blood, Sir Francis Bacon (1561–1626) had written great treatises on the nature of science, and three pioneer astronomers and physicists—Galileo Galilei (1564–1642), Johannes Kepler (1571–1630), and Sir Isaac Newton (1642–1727)—had laid the basis for heavenly cosmology and earthly physics. The work of these scientists of the sixteenth and seventeenth centuries was for the most part grounded on careful observations and experimentation rather than on preconceived notions of how the world ought to work. Galileo, Kepler, and Newton discovered that complex phenomena, such as the motions of celestial and earthbound bodies as well as gravitation, obeyed precise rules that could be expressed mathematically. This was the beginning of serious attempts to reduce all natural processes to scientific—and ultimately mathematical—statements. Conclusions were based increasingly on natural processes and less and less on metaphysical concepts.

Suddenly all problems relating to nature seemed to be approachable by the unfettered human mind, and a sense of freshness and freedom swept over Western intellectuals. In 1664 Henry Power, Dr. of Physick, described how it felt to be part of this great intellectual revolution: "This is an age wherein (me-thinks) Philosophy comes in with the Spring-tide. . . . Me-thinks I see how all the old Rubbish must be thrown away, and the rotten Buildings be overthrown, and carried away with so powerful an Inundation". Heady stuff. And he was not burned at the stake, mainly because he was an Englishman, and the Anglican Church was far less powerful in Britain than the Catholic Church was on the Continent. Times were changing, as Henry Power realized.

Science was stimulated by other developments as well. Western Europeans were voyaging around the globe, encountering new cultures and discovering hundreds of new species of plants and animals. The Royal Society of London, founded in 1662, became a center for scholars of Western Europe who were seeking a deeper knowledge of the natural world. Its publications served as a vehicle for making scientific knowledge available to all who might be interested. Universities throughout Europe were increasing in number and quality, and there was a slow but deliberate change in attitude about what was appropriate intellectual activity.

Rational versus Romantic Thinking

The triumph of rationalism in astronomy, mechanics, and anatomy spilled over into other human endeavors to such a degree that by the eighteenth century, Western civilization had entered a period known as the Age of Reason or the Enlightenment. The newly freed human mind no longer looked to the Bible or to the authority of ancient scholars for answers. Instead, it sought to study the natural and human worlds through the powers of logic and observation. Above all else, supernatural explanations were to be rejected. It became widely accepted among intellectuals that the universe operated according to natural laws, which could be discovered through scientific procedures and expressed mathematically. The knowledge gained could lead to great improvements in the human condition.

This is not to say that scholars of the Enlightenment uniformly rejected God or religion. Quite the contrary; they were usually pious Christians who believed that science was a useful tool for understanding God's work. According to the deistic world-view of the time, God could be known in two ways. One was to study "*the Word*," that is, the Bible, which was still accepted by the Church and nearly everybody else in the Judeo-Christian tradition as having emanated from God. The other was to study "*the Work*," that is, what He had wrought through creation. Scientists such as Galileo and Bacon were well aware of the many difficulties in interpreting what was said in the *Bible*, and they hoped that the "*book*" of nature might help to solve those puzzles. Interpreting nature, therefore, became a companion activity to the theologians interpreting the *Bible*. Together these pursuits would lead closer to truth—or so it was hoped.

Not surprisingly, the *Age of Reason* elicited a counter-reaction. The widespread attempt to "*be scientific*" in the eighteenth century was largely rejected by nineteenth-century intellectuals of the Romantic movement, who viewed the Enlightenment world as cold, heartless, and confining to the human spirit. In all this rationality, they asked, where was a place for inspiration, imagination, emotion, spiritual aspirations, self-expression, individual creativity, mystery, revelation, and tradition? Copernicus and his followers had demoted Earth from the hub of creation to a minor planet orbiting a minor star. The great eighteenth-century classifier of plants and animals, Carolus Linnaeus, had demoted humankind from the center of creation to a member of the Primate order, where a man or woman became just another ape. The Romantics accused science of finding no noble purpose for human life, overlooking the fact that finding purpose or creating meaning is not a proper assignment for science.

Human thought is far too complex to be divided strictly into these two discrete patterns—rational and romantic. Nevertheless, the pattern fairly typical among intellectuals of the Enlightment of the eighteenth and nineteenth centuries still characterizes much of our thinking in the modern world. During that time the pendulum swung noticeably toward the rational, critical, empirical, mechanistic, material, impersonal, skeptical, and reductionist mode. According to this ideal, unbiased observations and experiments produced data that others could confirm, and supernatural beliefs unsusceptible to scientific procedures were to be rejected. The world was a fit object of critical study, and not only was great progress possible in understanding it, but the products of science would inevitably increase the well-being of humanity. This philosophy—one might even call it a mission statement—is one that most scientists accept today.

But in some areas of human life, such as love, friendship, religion, the arts, and literature, a romantic way of thinking takes precedence over rational thought. Again, allowing for much fuzziness, one might characterize this kind of thinking as romantic, creative, emotional, humanistic, spiritual, personal, mystical, and holistic. But in no major undertaking does one mode of thought or action suffice. The design of an automobile that actually works is heavily weighted toward the operations of the rational mind; but if one hopes to sell the cars one manufactures, then style, beauty, and history must be given their due. The nurturing of a child requires both a romantic and a rational mind-set. Love and individual attention in a warm and supportive family situation are essential, but so is scientific information about nutrition and medicine for maintaining the child's health.

It is important to note also that intellectual debates between rationalists and romantics in the eighteenth and nineteenth centuries involved a mere fraction of society in a small corner of the world. Most human beings during that time lived, acted, and believed as human beings always had, and as most human beings do today. Moreover, we must not make the mistake of assuming that these two general patterns of thought first crystallized during the Enlightenment and Romantic periods. The Greek philosophers in Ionia were surely enlightened and they were just as surely capable of thinking in the romantic mode as well. Today, as then, the two points of view are different and often incompatible, but the consequences of that incompatibility are rarely severe. The human population is not divided into two hostile camps; everyone uses both patterns of thought, depending on what is being thought about.

The scientists make use of the romantic mode in at least 90 percent of their nonprofessional activities. The people they marry; the food they eat; the art, music, and literature they create or appreciate; the religious practices they follow; the recreation they enjoy; the pets they choose; and the clothes they wear are not selected on the basis of cold, calculating, impersonal data. And we can be thankful for that. On the other hand, few scientists I know—or anyone else, for that matter—would step off the curb into fast-moving traffic and expect a personal God to intervene and prevent their change from three dimensions to two.

Because all human beings are capable of these divergent thought patterns, one rational and the other romantic, there are no easy solutions for what to do with the bones of *Kennewick Man* or how to view the conditions of the *women of Afghanistan* under the control of the Taliban—or whether to teach *creationism* or *evolution* in the schools. There is no one acceptable answer to these social decisions, because different groups adhere, sometimes fiercely, to different answers.

2

DIVERSITY OF LIFE

Western civilization has developed two major ways of accounting for the diverse species of creatures alive today. The oldest, and probably the one most widely accepted in the United States, is based on a literal interpretation of Genesis, the first book in the Judeo-Christian *Bible*, which dates to a few centuries b.c. This is the familiar story of God's creation of the universe, our solar system with its sun and orbiting planets, and living creatures—all in six days. Strict creationists date these events to no more than 10,000 years ago. According to this account, a second dramatic influence on living creatures occurred when essentially all life, except for one or a few pairs of each kind or species, was destroyed by a worldwide flood. The few survivors were those taken aboard the Ark by Noah. All present-day life, therefore, is descended from the passengers on the Ark. The date of the flood is estimated to be in the third millennium b.c., possibly 2350 b.c. Creationists believe that all species have remained essentially the same as when they were created.

The entire evidence for the *creationists'* explanation for life on earth is found in Genesis. There is no independent confirmation of this account of creation from biology or geology, and, in fact, a tremendous body of information from these sciences argues that the creationists' stories cannot be correct. Creationists accept divine creation solely on the basis of their belief in a supernatural God, combined with an unshakable faith in the inerrancy of the Bible.

The second way of explaining the diversity of life is evolution. The modern version of this concept began its vigorous growth with Charles Darwin in 1859 and has expanded during the past century and a half. It is so important in our understanding of life that it has been said that nothing in biology makes sense without it. In contrast to the creationists' explanations that involve the supernatural, Darwin proposed that the concept of evolution must be based entirely on natural phenomena. An exclusion of supernatural things and processes is basic to the scientific approach that has given us the astonishing products of technology, modern medicine, a much more productive agricultural system, and, indeed, *modern civilization*.

And for the inquiring mind, we now have an understanding of the natural world that is, itself, a thing of beauty. The seeming chaos of our world and much of the cosmos is being replaced by an understanding based on a few overarching principles. One of the more basic of these is the absolute dependence of all life on the sun. Life is a consequence of the interactions of molecules in thousands of different chemical reactions. Some of these reactions release energy, but the net result is that energy from outside an organism is required for the chemical processes of life, which consist of the synthesis of simple molecules mainly into complex proteins, fats, carbohydrates, and nucleic acids. The ultimate source of this energy is the sun, but animals cannot bask in the sun and acquire the energy to keep their metabolic activities going. Most rely on green plants, which capture the radiant energy of the sun to combine simple molecules such as carbon dioxide, water, and a few salts into simple sugars. Species of the animal world are absolutely dependent on plants for food and or on other animals that eat plants. This food supplies the substance both for reassembling the plant molecules into animal molecules and for the energy necessary to do so.

The sun has other roles essential for life. Its radiant heat keeps much of the world at temperatures that permit water to be in the liquid state. This is critical since our bodies consist mainly of water, and the reactions in our cells occur in a watery environment. Heat from the sun evaporates the water that later comes back to earth as rain so necessary on the land masses that would otherwise be lifeless deserts. Rain also shapes much of the Earth's surface since it is one of the main factors in the erosion of mountains to plains and the carving of the surface into valleys by rivers and streams. Neither life nor the surface of the earth as we know it could exist without liquid water.

Vast quantities of the sun's energy from long ago are stored in deposits of coal, petroleum, and natural gas that can be used directly or indirectly in the generation of electricity for our technology, transportation, and heat and light. These natural resources, finite and precious, are in a sense fossil sunbeams. In so many ways the sun is the dynamo that drives our Earth. Its powerful role could not have been fully understood in ancient times, but, nevertheless, many early cultures worshipped the sun, often as their most important deity.

Once life appeared on Earth, that it would evolve was almost axiomatic. The simple reason is that the central property of life is its ability to reproduce, so an ever-increasing amount of life requires an ever-increasing quantity of resources. Apart from sunlight, all the resources required for life are finite, including a place to live, which is the most limiting resource for both plants and animals. The continents and the oceans may vary in size over geological time, but the sum cannot increase. The chemical substances that make up living bodies were always available not because they were unlimited but because they were cycled. Green plants take in carbon dioxide from the air along with salts and water from the soil and combine them to synthesize the chemicals that form their structures and that supply them with the chemical energy required for their life. Animals, by contrast, require oxygen from the air, water, and complex organic molecules in their food, which originate in green plants. The molecules involved enter organisms and are constantly returned to the environment as waste products and when the organism—plant or animal—dies. Molecules are borrowed for a life but not destroyed.

By Darwin's time, it was understood that the resources of the environment are limited: a single acre could not support a stand of trees, a herd of cattle, or a flock of sheep of infinite size. In 1798 Thomas Robert Malthus (1766–1834), an English economist and demographer, had suggested that human beings faced similar restrictions: populations of infinite size could not be supported by a finite environment. Charles Darwin read Malthus's work carefully and claimed that it provided the clue for how evolution might occur. It was also common knowledge in Darwin's day that a single plant in its lifetime produces a vast number of seeds, yet, by and large, the number of flowers in a field or trees in a forest remains fairly constant from year to year. Animals also have the ability to produce far more offspring than there are parents; yet the population size of any species seems to fluctuate around a mean. There was little evidence to suggest that species of animals and plants somehow restrict their own reproduction in order not to exceed the carrying capacity of the environment. Quite the contrary: many more offspring are produced than the environment can support, and nearly all the offspring of plants or animals die before reaching maturity—the world is simply not big enough for everyone.

Could there be qualitative differences between the few that survive and the many that do not? Darwin speculated that there might be. It was common knowledge that the offspring produced by two parents can differ in their characteristics. If some of these characteristics are inheritable, Darwin suggested, therein lies a mechanism for evolution. It seemed likely that some characteristics might do a better job than others in allowing the individual with them to survive and reproduce. The environment would select the better-adapted individuals, analogous to the way a breeder selects individuals with desirable traits and culls those with undesirable traits. Darwin called this process "*natural selection*." Continued for hundreds of generations, natural selection might produce a population so different from the original one that it could be considered a new species; and with much further evolution, it would become a new genus, then a new family, then an order, a class, and finally a phylum.

The three components of Darwin's suggestion for how evolution could occur—a finite environment, *genetic variation* among individuals, and natural selection—are all processes that can be studied in nature by anyone. The data that Darwin and later evolutionists obtained were all based on observations and experiments. Even before Darwin's time, geologists had observed the great thickness of the sedimentary layers of the Earth's crust and concluded that the Earth was very, very old. Today, data obtained from measurements of radioactive decay set the origin of the Earth at about 4.5 billion years ago and the origin of life about 1 billion years later. The scientific evidence does not support a young Earth, contrary to the claims of the creationists. So also the successive layers of sedimentary rocks preserve a diary of previous inhabitants in the form of the fossil record, which has been interpreted as showing the changes of species into new species. The paleontological data provide no confirmation for the flxity of the species that the creationists claim.

The *usefulness* of any scientific theory is its ability to explain a variety of otherwise inexplicable observations. The greater the extent to which it can do this, the more likely is it to be correct. The *theory of evolution* provided a naturalistic explanation for many puzzles in comparative anatomy; the classification of organisms; embryonic development; physiology; genetic makeup; the structure and physiology of cells; the biochemical reactions in microorganisms, plants, and animals; the geographic distribution of species; and observations in the paleontological record. To date, there have been no confirmed observations or experiments that falsify the theory of evolution.

Can we say, therefore, that evolution or any major concept is "*true*" or that it is a "*fact*"? Scientists prefer not to use those terms, based on their experience that all scientific "truths" and "facts" relating to major concepts have been modified and improved with time. For example, long ago it was believed that all matter was composed of four basic elements—earth, water, fire, and air. The notion that a small number of substances combine to form all other substances was a powerful idea that started in ancient Greece and has proved to be correct, but the substances first specified as elements proved to be incorrect. Later scientists hypothesized that all matter is composed of similar indivisible structures—atoms. In the nineteenth century, about 100 basic kinds of atoms had been discovered or predicted to exist, and these were thought to be the basic building blocks of all matter. Then in the twentieth century the atoms themselves were discovered to be composed of electrons, neutrons, protons, quarks, and ever smaller particles. One current theory hypothesizes that the basic building blocks of matter are superstrings—one-dimensional, vibrating massless strings only 10_{33} centimeters long.

A useful way to look at the history of science is that the earlier statements were not wrong but merely incomplete. Each statement represented the best that could be said with the data available; when better data became available, the statements were upgraded. Nevertheless, some theories are so well established that they can be considered "*true*" in the common sense of tnat word. The theory of gravity is one of them: if a heavy object is released from the hand, it falls; we can be sure of that—as long as we remain on the Earth. In a spacecraft, heavy objects will float in air.

In the 1940s, noted English biologist Julian Huxley expressed the opinion that the theory of evolution was as well established as the theory of gravitation. No new data or experiments have changed that opinion. Nevertheless, the theory continues to be augmented with new data and interpretations, and—of the greatest importance—it suggests new lines of research and discovery. If Darwin could pay us a visit today, he would probably be delighted to find his basic ideas still useful, but he would probably be astonished at the progress made since 1859, much of it in genetics and molecular biology—fields that didn't exist during his lifetime.

Patterns of Thought

The differing explanations that *creationists* and *evolutionists* provide to explain life over the geological ages illustrate the two alternative ways human beings think about their world, their hopes, and their lives. One approach is rational, demanding data and logic, while the other way is more romantic, involving emotion, faith, and personal preference. The rational mode is more likely to be observed in the sciences, technology, business, law, medicine, government, and education. The *romantic mode* is more likely to be found in music, art, literature, religion, interpersonal relationships, and lifestyles. Most of the time both patterns of thought are involved in decision making. A desire to improve the lives of those living in poverty originates in the romantic mode, but if that desire is to bear fruit, rational thought must be applied to implementation of the goal.

The disputes between the creationists and scientists cannot be considered scientific, since only one side deals with science. Nor can they be considered religious disputes, since the theories of science have nothing to say about gods or other *supernatural phenomena* that cannot be studied by scientific procedures. The disputes are best thought of as political disagreements, not scientific ones. In instances where two points of view exist—each held firmly—an effort could be made by each side to maintain its own views (both will) and ignore the other.

Both sides will have to make adjustments—not in their beliefs but in their actions. The creationists can keep their faith, but they must refrain from trying to force their ideas on the science curriculum in the classroom. Professional scientists need to initiate a full discussion in our society of what science is and is not and how scientists know what they claim to know. A large percentage of Americans remains ignorant of how scientists discover new knowledge and unite that knowledge into conceptual schemes. For people not thoroughly familiar with biology and geology, the relevance of the evidence presented for evolution is not obvious without a good background in these sciences.

Little in modern science is readily understandable without considerable specialized education. Is it obvious to nonphysicists why scientists accept that atoms are made up of many different kinds of particles, how scientists know that the rotation of the Earth rather than a circling sun causes night and day, why ice floats, how the sun's energy drives the workings of our bodies, or how it was determined that DNA is the hereditary material? All of us, including scientists, are amateurs when it comes to understanding the cutting-edge developments in fields other than our own. Few biologists can comprehend discoveries in theoretical physics, and most physicists and many biologists have little understanding of the data of modern evolutionary biology. We have to accept that everyone is ignorant of almost everything.

Our acceptance of the scientists' word about so many things in nature rests on the belief that those who devote their lives to the study of a subject are more likely than others to have reliable information. And we know that when scientists reach an erroneous conclusion, the errors and omissions leading to that conclusion will eventually be detected as other scientists in the field repeat the experiments or observations and extend the analysis. In other words, the system is self-correcting, and that is its greatest strength. If equally competent scientists with equally effective equipment anywhere in the world perform the same experiments or make the same kinds of observations, the results are likely to be similar. If the outcomes differ, repetitions of the work will reveal the sources of the differences. Science advances by the retention of the provable and the elimination of the falsifiable. Though politically and culturally we are far from seeing ourselves as one world, scientifically we come very close.

Professional scientists need to become informed enough to discuss, as citizens repeatedly ask them to, the difference between creationism based on faith and evolution based on confirmable evidence. And religious leaders should support the scientists in this effort. Many theologians deplore the erroneous statements and the political activities of creationists. They could make clear to their congregations that acceptance of evolution does not mean denial of religion, but it does mean regarding the P *version of creation* as *metaphor*, not science. And they should inform their parishioners that many of the major religions in the Western world and their leaders accept evolution as the best explanation for the organic diversity of the present and the past.

Teaching Science

Concerned parents and other citizens should do their part by working with school boards and teachers to make sure that science is taught as it has been developed by scientists and not according to the views of those who wish to discount it. The educational system, which involves parents, schools, and society, does not give adequate consideration to developing both the romantic and rational aspects of the mind. The world of young children has a rich romantic and supernatural component, with influences from children's books, motion pictures, churches, and parents. Little consideration is given to the processes that require evidence and critical thought for reaching conclusions. Much anguish and conflict might be avoided later if children were taught in their formative years that both romantic and rational thought patterns are valuable for effective interrelations of the individuals of any society.

Children in kindergarten and elementary school are highly receptive to topics in science, especially if those topics pertain to the world they know. Through simple observations and experiments, children learn that that they can answer some questions about nature themselves. Unfortunately, teachers in these beginning grades rarely are knowledgeable enough about the natural sciences and critical thinking to lead their students in explorations and analyses that will help develop their rational patterns of thought. Correcting this deficiency will require a major overhaul of the existing educational system—both in the ways teachers are educated and in the opportunities for children to learn.

Few colleges and universities offer science courses specifically designed for prospective schoolteachers. Science courses in higher education are designed primarily for students who plan to be professional scientists, engineers, or physicians. The curriculum for these students tends to be rigorous, selective, and restrictive. But such courses alone do not provide an adequate preparation for teachers. Prospective teachers should have a solid background in the subjects they will teach, but they also need courses to help them teach science in ways appropriate for elementary schools, middle schools, or high schools. They must be familiar with all fields of science and how they relate to the lives of human beings. Students should learn not only the basic principles of all sciences but also topics such as the differences between science and other disciplines. Critical thinking and what counts as evidence should be understood. There should always be a deep concern for the relationship between the sciences and public policy and the conditions of the environment that determine human welfare. And for all levels of education, and appropriate for the ages of the students, science should consider the concerns and interests of people.

Many high school science teachers actually teach creationism, which can only be regarded as a lack of professionalism and a *violation of law*. But this may not always be the teacher's fault. Few college and university science courses deal with the creation-versus-evolution controversy explicitly, such that students, including those who will be teachers, can understand the issues and the data and be able to make sound judgments. Again, scientists must accept much of the responsibility for this situation.

Moving Forward

The *evolution-versus-creationism* debate can be viewed as a conflict between two incompatible paradigms, the romantic and the rational. For centuries humankind has spoken of the affairs of the head and of the heart as different realms. Romantically they are; rationally they are not. Both scientists and humanists require both a thinking head and a pumping heart. The prognosis for those who choose one or the other is poor. The challenge is to employ the metaphorical heart and the metaphorical head in ways that emphasize the strengths and limitations of each.

The feeling of awe that comes from understanding the natural world and the unifying principles that control the interactions of all matter can be a deep religious experience. One need only think about the long history of the great mountain ranges to be awed by the mighty forces of the crustal plates crashing against one another, pushing the surface upward to form mountains, and the power of the wind and rain that sculpted them. And every living creature reminds one not only of its long history going back to the origin of life itself but of the extraordinary structures and functions that permit its survival in what can be an unkind world. One can experience a profound sense of wonder in being a part of a knowable world.

Whether or not God is the driving force behind that world is not for any scientist, speaking as a scientist, to say. Scientists as scientists can deal only with phenomena that can be studied, and this does not include God. As individuals, however, they can accept that there is a God—a position that is based on belief, not scientific evidence. According to author and editor Gregg Easterbrook (1997), there is a growing feeling among many scientists and religious leaders that accommodation and even mutual support of their positions may be the mode of the future: "Perhaps the fact that the two schools of thought have so often been at each other's throats stems from mutual recognition of their linked destinies, and their joint commitment to the idea that the truth is out there. Rather than being driven ever farther apart, tomorrow's scientist and theologian may seek each other's solace". Harvard biologist E. O. Wilson (1998) also predicts that science and religion are moving toward synthesis.

Physicist, cosmologist, and Nobel laureate Charles H. Townes expands upon this point of view (1995):

The ever-increasing success of science has posed many challenges and conflicts for religion—conflicts which are resolved in individual lives in a variety of ways. Some accept both religion and science as dealing with quite different matters by different methods, and thus separate them so widely in their thinking that no direct confrontation is possible. Some repair rather completely to the camp of science or of religion and regard the other as ultimately of little importance, if not downright harmful. To me science and religion are both universal, and basically very similar. In fact, to make the argument clear, we should like to adopt the rather extreme point of view that their differences are largely superficial, and that the two become almost indistinguishable if we look at the real nature of each. . . .

The goal of science is to discover the order in the universe, and to understand through it the things we sense around us, and even man himself. This order we express as scientific principles or laws, striving to state them in the simplest and yet most inclusive ways. The goal of religion may be stated, as an understanding (and hence acceptance) of the purpose and meaning of our universe and how we fit into it. Most religions see a unifying and inclusive origin of meaning, and this *purposeful force* we call *God*. . . .

The essential role of faith in religion is so well known that it is usually taken as characteristic of religion, and as distinguishing religion from science. But faith is essential to science too, although we do not so generally recognize the basic need and nature of faith in science.

Faith is necessary for the scientist to even get started, and deep faith necessary for him to carry out his tougher tasks. Why? Because he must be personally committed to the belief that there is order in the universe and that the human mind—in fact his own mind—has a good chance of understanding this order. Without this belief, there would be little point in intense effort to try to understand a presumably disorderly or incomprehensible world. Such a world would take us back to the days of superstition, when man thought capricious forces manipulated his universe. In fact, it is just this faith in an orderly universe, understandable to man, which allowed the basic change from an age of superstition to an age of science, and has made possible our scientific progress. . . .

Finally, if science and religion are so broadly similar, and not arbitrarily limited in their domains, they should at some time clearly converge. This confluence is inevitable. For they both represent man's efforts to understand his universe and must ultimately be dealing with the same substance. . . . But converge they must, and through this should come new strength for both.

It is hard to imagine a synthesis occurring anytime in the near future because science and religion still use very different thought patterns— one based on evidence and the other on belief. Rather than predicting a synthesis of science and religion, it may be more useful to regard them as coequals with different domains. Science attempts to understand how the natural world works, and this knowledge not only feeds our natural curiosity but also provides us with astonishing power to live and to change our lives. But for most people this knowledge is not enough; the question "What does it all mean?" is equally important. In a general way we can say that religion deals with questions of meaning as well as with standards of conduct. But these meanings and standards are assigned by human beings, not derived from what science tells us about the mechanisms of life. There is one science for the entire world, but innumerable and often conflicting answers to questions about the purpose and meaning of things. "What is the purpose of human life?" has answers that vary from one human being to another and even vary at different times for the same individual. Herein lies a gap that is so far unbridged. Whereas science is discovered, meaning is assigned.

From this standpoint, science and religion need not be seen in conflict because their domains are distinct. We should recognize that the first is preeminent in providing knowledge of the natural world that can be exciting and awe-inspiring as well as provide for modern medicine and the comforts of modern life. Those comforts are not to be enjoyed unless each society can agree on a moral code of behaviour so necessary for civilized life.

However, science and creationism come into serious conflict when the politically active antievolution creationists, or creation scientists as they sought to be called until recently, campaign to have their brand of creationism taught in the schools alongside evolution or to be given "*equal time*." This subset of creationists are not scientists; they do not see new information in laboratory studies or field explorations to test their hypothesis. They ignore the conclusions of scientists who over the past two centuries have provided a body of confirmable evidence that leads to the conclusion that, beyond all reasonable doubt, evolution has been the dominant phenomenon in life over the ages. Creationists have every right to ignore all this and believe whatever suits them, but as the highest courts of the nation and some states have ruled, creationism is not science and cannot be taught as science in the nation's public schools.

Religion has had a *checkered influence* during human history; it has been responsible for much bloodshed and misery in the past as well as today in many parts of the world. This is less a function of religion itself and more the interpretation of religion as sanctioning what individuals, cultures, and nations wish to do. Battling nations each proclaim that *God* is on their side, which suggests a robust polytheism. "Go ye into all the world and preach the gospel to every nation" has been read as a command to destroy the cultures of many native societies in North and South America, Australia, sub-Saharan Africa, and islands throughout the world. And much of this strife among different peoples has been supported by discoveries in science and the conversion of that knowledge into instruments of personal and mass destruction. Science as a way of knowing nature is neutral, but its power can be used for infinite good or infinite evil. The choice is ours.

Throughout the ages there have always been a few who, without abandoning their religion, did forgo the blind acceptance of dogma and superstition. They undertook the extraordinarily difficult, lonely, and frequently dangerous path of using their unfettered minds for rational inquiry. It is these individuals who have given us the modern world and the possibility of truly great improvement of the human condition. They have replaced the primitive view of nature as chaotic, mysterious, and often threatening with a view of the universe and life as responding in patterns that are precise, beautiful, and awe-inspiring. Beyond giving pleasure to the inquisitive, analytical mind, this progress in understanding provides previously unimagined ways to feed the hungry, heal the sick, and lessen toil. Lives are poorly lived when they look out upon a cold, hostile, inscrutable world; lives are enhanced when they look out upon a world with an appreciation of its beauty and order and its suitability as a warm and friendly home. It matters little for the great moral and ethical questions facing humanity whether or not the human brain and mind are consequences of random events in evolution, though scientists are convinced they are. However, it matters a great deal that we use our brains and minds honestly, humanely, intensively, and effectively to preserve and improve the world for ourselves and for the generations that follow.

3

Inheritance and Speciation

Darwin's magisterial assembling of the *indirect evidence* for evolution was convincing to many scientists almost from the publication of the Origin, and by the end of a decade to nearly all. However, his hypothesis for the mechanism of evolutionary change, namely, inherited variations acted upon by natural selection in a finite environment, was not so convincing. A major reason for this was the near total lack of information in Darwin's time about the origin of variation and the workings of inheritance.

Rules of Inheritance

In the last half of the nineteenth century it was not clear whether or not inheritance is a constant and repeatable phenomenon. Some observations suggested that there might be rules of some sort for inheritance, but equally persuasive evidence suggested a fickleness at best. It was obvious, of course, that the offspring of human beings were human beings and that this principle applied to all known life—the species of animals and plants "*breed true*." It was equally obvious that children of the same human parents were far from identical except in the rare cases of identical twins. Sometimes parental characteristics seemed to be inherited, but in other cases, not. One of the most confusing of all was that a parental feature would not be expressed in the children but would reappear in the grandchildren. And then there was that most astonishing difference of all: offspring of the same parents can be either *males* or *females*, which differ greatly in structure, physiology, behaviour, and reproduction. In most species that reproduce sexually, about half the offspring are females and the others males. It is obviously convenient that approximately equal numbers of males and females are born, but convenience is not a mechanism—so what could possibly account for this phenomenon?

Darwin's theory required a pool of inherited variations among individuals of the same species from which natural selection could choose those that were more fit for survival and reproduction. When Darwin published the *Origin*, he believed that inheritance must have a definite biological basis and that, whatever it was, it had to provide offspring like the parents but with *minor variations*. There were no data to explain how this was possible during the last half of the nineteenth century. In subsequent editions of the Origin Darwin was driven more and more to the belief that the variations he saw in nature were due in some degree to the direct action of the environment. This had been the suggestion of Lamarck back at the beginning of the century, and by the century's end a belief in *Lamarckism* had become common among Darwin's contemporaries.

The modern understanding of inherited variations began in 1900, more than a decade after Darwin's death. In that year, experiments on inheritance in garden peas conducted in the 1860s by an Austrian monk and amateur naturalist, Gregor Mendel, were rediscovered and made known to the scientific world. Mendel had crossed varieties of garden peas that differed in characteristics such as the shape and colour of the seeds or the length of the stems. He observed patterns of inheritance in the second and subsequent generations of peas, and he made sense of these patterns by recognizing that each characteristic he studied came in two versions. Today, we would call the versions of a given gene responsible for this kind of *variation alleles*. For example, a gene for seed colour in Mendel's peas had two alleles—one for yellow and the other for green. When both alleles were green, the seeds in the first and second generations were green. When both alleles were yellow, the seeds in the first and second generations were yellow. But they were also yellow if one allele was yellow and the other green. That is, the yellow allele was dominant, and the green one was recessive. Today we would say that when together the yellow allele was "*expressed*," while the green one was not.

When an organism has two different alleles for a given characteristic, say, yellow and green, the organism is heterozygous for that characteristic. When an organism has two *identical alleles* for a given characteristic, we say the organism is *homozygous* for that characteristic. In Mendel's experiments, when two pea plants that were *heterozygous* for seed colour were crossed, three-quarters of the offspring produced yellow seeds and one-quarter produced green seeds. The green seeds were just as green as they would be in a pure-breeding green pea plant. They were not a yellowy green. This meant that the two alleles, for green and yellow, were not affected by their coexistence in a heterozygote; they remained particulate, that is, separate.

Mendel established the important principle that there are strict rules for inheritance that can be expressed in *mathematical ratios*. In the case of many characteristics, if the kinds of alleles in the parents are known, the percentages of the kinds of offspring can be predicted. Furthermore, the alleles seemed to be stable—Mendel did not observe any new variations (what we would call mutations). But were these discoveries restricted to garden peas, or did they apply to other organisms? Through intense experimentation with many different species of plants, other geneticists discovered that with minor variations, Mendel's rules held for other sexually reproducing species as well.

The next major advance in understanding genetic mechanisms came in the early 1900s, when Thomas Hunt Morgan and his students at Columbia University began to study inheritance in the small fruit fly Drosophila melanogaster. This species is commonly found hovering above fruit, and the Morgan group collected individuals for genetic studies from ripe bananas placed on the windowsill outside their laboratory. *Drosophila* proved easy and inexpensive to raise in the laboratory, and its generation time was short—less than two weeks. This meant that the results of crosses could be determined quickly, compared with crosses of peas, mice, or human beings. It was also possible to make certain types of specific crosses with *Drosophila*: mothers to sons, or fathers to daughters, for example. Such incestuous crosses are important in detecting *recessive alleles* that are carried in the heterozygous state and hence are masked by the *dominant alleles*.

With these techniques, the Morgan group discovered many alleles for a given characteristic in fruit flies. Using these alleles as *markers*, they rapidly established that genes are parts of chromosomes. Previous work had established that the *chromosomes*, which are found in the nucleus of cells, are present in pairs—*Drosophila* has eight chromosomes in four pairs, whereas humans have 46 chromosomes in 23 pairs. Each gene is located at a fixed place in a chromosome, so in an organism every gene is represented twice—one on each chromosome of a pair. However, the gene may have one version (say, a yellow-seed allele) on one chromosome and a variant version (a green-seed allele) on the other chromosome. Morgan and his colleagues discovered that unlike the case with seed colour in garden peas, most features of the fly are affected by many different genes, not just one, and a single gene usually affects several different structures. Thus, tracing the patterns of inheritance of complex structures proved more difficult than Mendelian genetics had initially led geneticists to believe.

In other experiments Morgan's group discovered that the cells that form the eggs or sperm undergo complex changes that randomly reshuffle the chromosomes with their genes. Thus, each egg or sperm gets a set of reshuffled chromosomes that is different from the chromosome of the parent and from that of every other egg or sperm. When egg chromosomes of one individual combine with the sperm chromosomes of another individual during *fertilization*, the result is a huge amount of variability in the offspring.

But *sexual reproduction*, with its associated recombination of genes, is not the only source of the variation on which natural selection can act. Another source is the formation of entirely new alleles when a gene mutates. The rate of mutation is generally very low under natural conditions, but early experiments showed that the rate can be increased by external factors such as exposure to X-rays, high temperatures, or mustard gas. The list of mutation-causing agents (mutagens) that we know today is far longer. But whereas the rate of mutation can vary depending on the stimulus chosen, Morgan and his group found that the kinds of new mutations were not related specifically to the stimulus—the same sorts of mutations occurred no matter what the stimulus.

This discovery of the random nature of mutation was very important for evolutionary theory. It suggested that mutations favouring a thick fur coat (in a mammal) are no more or less likely to occur in the Arctic than mutations favouring a thin fur coat. That is, mutations are not caused by the environment or by the specific needs of the organism. What causes polar bears in the Arctic to have thick fur coats rather than thin ones is natural selection acting on this random tendency to produce thicker or thinner coats. *Gene mutation* is constantly producing new alleles that are then screened by natural selection for their adaptability to a given environment; if they are of advantage to individuals of the species, they will eventually be incorporated into the species' genetic makeup because more individuals with the *advantageous mutation* will survive and reproduce than individuals without it.

When geneticists studied the alleles of individuals from populations in the wild, they discovered that these individuals were heterozygous for many of their genes. Such abundant heterozygosity, plus the reshuffling of genes when eggs and sperm cells are formed, followed by fertilization between genetically different eggs and sperm, provides for tremendous *genetic variation* among offspring. So much, in fact, that except for identical twins, every human being probably differs from every other human being who is now alive or who has ever lived.

Inheritance—the rules for the transmission of genes from parents to offspring—had become an extraordinarily exact science by *World War II*. At that point the interests of geneticists turned elsewhere: to the questions of what genes actually are and how they act. Experiments to determine the chemical structure of genes began in the 1930s, and by the 1950s James Watson, Francis Crick, and others had discovered that genes are strings of nucleotides that are part of the nucleic acid molecule *deoxyribonucleic acid*, or *DNA*. The main function of genes is to produce protein molecules.

Within a gene the four *nucleotides* (*adenine*, *thymine*, *cytosine*, and *guanine*, abbreviated A, T, C, and G) are combined in functional groups of triplets. Each *triplet* is a template for the formation of a codon of messenger RNA (mRNA), which is composed of three of the four nucleotides: U, A, G, and C. Each codon in turn determines a specific amino acid that will be added to a growing chain of amino acids. Table 3.1 illustrates the translation of DNA to mRNA for a short section of a gene with the code GCA GGT TAC GTC. There are 64 possible combinations of messenger RNA codons, each responsible for inserting a specific amino acid or for stopping synthesis. This chain will eventually fold in complex ways to become a protein molecule. Many, indeed most, of the proteins that cells make are enzymes, which facilitate chemical reactions within the organism. All the biochemical events that occur within cells are the result of the activities of enzymes and other proteins. Thus, the mighty DNA with its many genes is the boss, but it is not the workforce. DNA can do only two things: it can replicate itself, which it does whenever a cell divides, and it can determine the specific proteins that cells make. It performs these functions in all organisms, from the simplest bacteria to the most complex mammals.

The *messenger RNA* codons are also the same in general structure in all species. This system is called the *genetic code*. Whenever a system as complex as the genetic code is found to be essentially the same in most species, we can be reasonably sure that it was established very early in the evolution of life and was conserved thereafter. Its

Table 3.1. The Synthesis of a Protein Molecule

If the DNA triplets are	*Messenger RNA codons will be*	*Amino acid added to the protein chain will be*
G	C	
C	G	arginine
A	U	
G	C	
G	C	proline
T	A	
T	A	
A	U	methionine
C	G	
G	C	
T	A	glutamine
C	G	

very complexity almost ensures that it will persist, because any new system would be in competition with an already working model. There is no direct way of testing this idea in fossil organisms, but it is a satisfying naturalistic explanation for the remarkable degree to which many cell structures and cell processes have been conserved and now characterize most living organisms.

Natural Selection

The inability of Charles Darwin or anyone else in the last half of the nineteenth century to demonstrate that *natural selection* operates in wild populations was a serious problem for understanding evolution. Scientists at that time generally accepted that evolution had occurred, but there was no agreement that natural selection acting on variation was the mechanism, as Darwin had proposed. If natural selection was the mechanism for evolutionary change, why did one not observe species becoming better adapted to their environment? After all, animal and plant breeders could mold one variety into another in just a few generations of selective crossing. Why did nature not act with equal speed? If it did, naturalists should be able to observe evolutionary changes within their lifetime. Yet none were apparent.

A satisfactory answer to this question did not come until well into the twentieth century. First, it had to be understood that natural selection does not convert a species into the best imaginable new population; all selection can do is produce a population that continues to survive and leave offspring. Once a species has reached that level of *adaptation*, the pressures for becoming better adapted decrease. Natural selection is for getting by, not for producing the best possible form of life. Any natural population that survives has already withstood every assault its environment could throw at it. Once a natural population has adapted to its environment, natural selection has the important role of eliminating individuals that may be abnormal or poorly adapted because of the chance combinations of genes they inherited from their parents. Thus not only does natural selection have a creative role in providing a mechanism for adapting to new environmental conditions, but it also has a cleansing role in maintaining the population by eliminating individuals with deleterious genes. This cleaning effect is the most important role of natural selection in a population in a stable environment. If the environment changes in a significant way, however, the population is no longer as well adapted, so it has to either adapt to the new conditions or become extinct. When the environment shifts, evolutionary changes may be more rapid, and a new species may be the result.

Once these relations were understood, how one could study selection became obvious: a population adapted to one environment could be confronted with a very different environment that would challenge the population in new ways. Any changes in the population would be good evidence that natural selection was operating. Classic studies demonstrating natural selection were done by two English naturalists, J. W. Tutt in the late nineteenth century and Bernard Kettlewell a half century later. The wings of the peppered moth have a complex pale and dark spotted pattern suggested by its name. In unpolluted rural areas where the trees have light-coloured lichens growing on their dark trunks, a resting moth becomes almost invisible because of its pale colouration. Lichens, however, are highly susceptible to airborne pollution, and a very different situation is found in industrialized regions. In the nineteenth century when industry began polluting the atmosphere, the lichens would die, leaving bare, dark tree trunks. A pale-coloured moth on a dark trunk would be a highly visible food source for a predatory bird. It was observed, however, that not many light-coloured moths inhabited industrial regions where nearly all of the individuals were very dark and hence protectively coloured—a phenomenon called *industrial melanism*. Regions with little or no air pollution, however, still had trees with the pale, lichen-covered trunks and the pale form of the moths. Genetic experiments showed that the colouration pattern is inherited. But was there any natural selection? Careful observations consisted of watching, hour after hour, the moths on tree trunks to see which colour type was caught by birds. It was found that the pale moths on dark trunks and dark moths on pale trunks were captured far more frequently than pale moths on pale trunks or dark moths on dark trunks.

In some areas where air pollution was later greatly reduced, lichens were found growing on the tree trunks once again. When this occurred, the dark form of the moth was replaced by the pale form. The observation that the evolutionary changes reverse when the environment returns to the prior condition is an important confirmation of the hypothesis that the changes are due to the levels of pollution. That is, if melanism is caused by increased pollution, a reduction in pollution should cause a decrease in melanism.

This pattern of change to the dark form and then reversal to the light form makes sense only if random mutations to the dark form are constantly occurring in the original populations of pale moths. Dark moths appearing in a rural woodland with lichen-covered tree trunks would be conspicuous to hungry birds. Natural selection in the form of those hungry birds would, therefore, eliminate the dark forms of the moths. Consequently, the moth population would remain almost entirely pale-coloured. The forces of natural selection would change radically, however, when the air started to become polluted and the lichens died, leaving the dark-coloured tree trunks exposed. The dark form would now hold the advantage, and the pale forms would be eliminated by natural selection. *Industrial melanism* is now known to occur in many species of moths from many parts of the industrial world, and it is known to decrease if pollution decreases.

Many natural populations besides moths show rapid evolutionary changes when they are confronted by a radically new environment. When *antibiotics* first became available a half century ago, many previously fatal diseases were easily controlled and cured by drugs such as *penicillin*. This drug was widely, even excessively, prescribed, and in just a few years an alarming situation developed. Strains of bacteria that were previously destroyed by penicillin developed resistance to the drug. New types of antibiotics were then developed; but in time, bacteria mutated and became resistant to the newer drugs as well. There are similar examples of populations of insect pests, for example, house flies, cockroaches, human head lice, and the fleas of dogs and cats, becoming resistant to the chemical pesticides developed to destroy them.

The development of resistance in bacteria and insect pests are forms of evolutionary change, and it is astonishing that they happen so rapidly. We normally think of evolution as an exceedingly slow process, and this is still true in a constant environment. But when a population is confronted by new and life-threatening environmental challenges, rapid change is possible. When a population of insects encounters a highly toxic pesticide in its environment, one which it has never encountered in the past, nearly all individuals are likely to be killed. But chance mutation in a few individuals may provide a low level of resistance—enough for these few to survive and reproduce. Since the resistant individuals are the only ones remaining to breed, their genes for resistance are transmitted to the next generation. Subsequent mutations of other genes could increase resistance and so increase the percentage of survivors to the point that the genetically resistant individuals become dominant in the population. Not until resistances in bacteria were studied did it become possible to determine whether gene mutations occurred randomly or whether a specific environmental challenge caused specific mutations. Experiments showed that the frequency of gene mutation leading to penicillin resistance in bacteria is the same whether or not penicillin is present in the culture medium.

Different *genes mutate* at different rates. On average, a given gene might mutate once in one individual every generation in a population of a hundred thousand to a million individuals. Many populations of plants and animals are at least that large, which means that in a given generation every gene of a species is mutating at least once in some individual. Since new mutations are almost always recessive, they will be masked by the dominant allele. However, these recessive alleles are carried along in the population. The tremendous amount of reshuffling of alleles that occurs when sperm and egg cells form and in fertilization results in some individuals that are homozygous for the recessive alleles. They will provide the genetically different individuals on which natural selection can act.

Rates of Evolution

The *fossil record* shows that the rate of evolution varies considerably. The data are rarely good enough to provide more than estimates. Some fossil remains of mammals from the end of the *Pleistocene*, about 10,000 years ago, are identical with living species. The many species of mammals and birds mummified by the ancient Egyptians 3,000 to 4,000 years ago are the same as present-day individuals. These data suggest that it takes longer than 10,000 years for one species to evolve into another. There are other ways to obtain estimates for rates of evolution. During the *Pleistocene* Ice Age the sea level was lowered due to the immense amount of water stored in ice sheets in the polar regions and in glaciers that covered the northern parts of the *Old* and *New Worlds*. England was then a peninsula of Europe. When the ice melted about 10,000 years ago, both England and the island of Jersey in the English Channel became isolated again. Today a few of the animals on Jersey are different enough from those on the European mainland to be classed as subspecies. Most, however, have remained the same. Other data suggest that it takes about 500,000 to 1,000,000 years for the evolution of a new species in birds and mammals.

Some general numbers can be given for the time interval from the beginning of one genus to the beginning of the next. Eight genera of fossil horses are now known that may form a lineage from *Hyracotherium* to *Equus*. *Hyracotherium* lived about 55 million years ago and Equus probably appeared about 1 million years ago. Therefore, if seven pre-*Equus* genera evolved in 54 million years, the average duration of each would be 7.7 million years.

The data for fossil vertebrates being among the best available, it is possible to estimate the time for the origins of the different classes of vertebrates. The oldest known class, the very primitive fishlike forms called *agnathans*, first appeared about 500 million years ago. It took about 100 million years for some advanced agnathans to evolve into the bony fishes. After another 50 million years some advanced fishes evolved into the amphibians, and 60 million years later the *first reptiles* appeared. After an additional 100 million years, the birds and mammals evolved from the reptiles. Thus ancestors of the mammals spent about 50 to 100 million years in each of the lower classes. Only a few species in a given class evolved into a higher class. For instance, after the *first amphibians* evolved they radiated into a large number of different species. Most became extinct; others persisted and evolved as the amphibians of today—mainly frogs, toads, and salamanders—while another kind of amphibian evolved into the primitive reptiles.

The immense duration of the vertebrate classes should be compared with the relatively short period for the major evolutionary radiation of orders of higher mammals—those with a placenta. Although the *first mammals* evolved from reptiles in the *Jurassic period*, it was not until the beginning of the Tertiary period that placental mammals began to evolve their great diversity: shrews, moles, horses, cattle, whales, dolphins, anteaters, elephants, lions, wolves, bats, plus many kinds that became extinct. Nearly all living mammals are placentals, and almost their entire history has taken place within the past 65 million years.

In contrast, a few genera have shown essentially no detectable change over millions of years. For example, the horseshoe crab, *Limulus*, so common on the Atlantic coast of the United States, has remained essentially unchanged for about 185 million years, and closely similar genera are known from 360 million years ago. The lineage of a *Cambrian* creature similar to the living species *Peripatus*—a fascinating animal with characteristics of both the annelid worms and arthropods—suggests that this species has persisted with little change from 570 million years ago to the present.

In recent years paleontologists have emphasized that evolution does not always occur at a slow, constant rate but is characterized by long periods of little or no change followed by (geologically) short periods of rapid change. For example, the orders of mammals evolved rapidly in the early *Tertiary*, after which they went through a period of slow evolutionary radiation within each order. This same phenomenon may apply to the evolution of species as well; that is, change may be relatively rapid as a new species evolves, followed by a long period during which the species remains essentially unchanged. This phenomenon was named "*punctuated equilibrium*" by Stephen Jay Gould and Niles Eldredge. It suggests that new species arise when the environmental conditions change and natural selection begins to alter the genetic makeup of the population to adapt better to the new *environmental pressures*. Once these adaptive changes have been made, there is little further evolution of the new species as long as the environment remains constant.

Speciation

New species can evolve in two major ways. First, a parent species can evolve into a single new species. Second, a parent species can evolve into two or more new species. In the first method the number of species remains the same; in the second it increases.

If the only way to produce new species had been evolution of one species into one other species over time, there would be only one species alive today. This model is obviously not adequate to explain the huge variety of living species we see in the world around us and in the specimens of the fossil record. Clearly, evolution also occurs by a single species diverging into two or more lineages that over time become different from both the ancestral population and each other. The original *gene pool* becomes two independent gene pools. But how does such a split occur?

By the middle of the twentieth century a convincing hypothesis was available to answer this question. If a single population becomes geographically divided, the subpopulations become subject to evolving in different ways in response to the environmental pressures they each encounter. This pattern of evolution is called *geographic speciation*. It is not the only mechanism for speciation, but it remains the one best documented. The accumulation of the relevant data for this hypothesis started in the nineteenth century, when traveling naturalists such as Darwin and Wallace began observing differences in populations that were seemingly of the same species but geographically separated. In

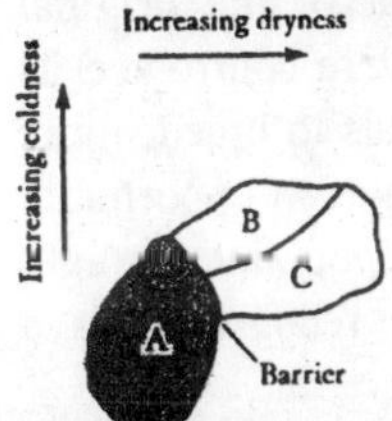

Time 1. Single species.
The population is restricted to Zone A.

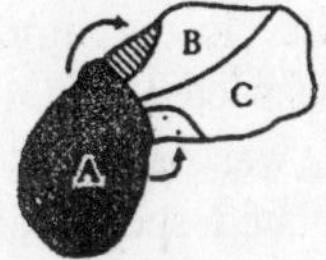

Time 2. Single species.
A few individuals migrate to Zone B and others to Zone C, where they are geographically isolated from the original population. Selection will promote the development of two new populations, one adapted to a cold, dry environment in B, the other to a warm, dry environment in C.

Time 3. Three subspecies.
Evolution through the selection of spontaneous mutations in the three isolated populations has reached the point where each zone has its own locally adapted population. These populations, though slightly different from one another, could still interchange genes should the geographical barriers between populations disappear. Consequently, they are subspecies, not yet true species.

Time 4. Three species.
The three isolated populations have diverged to the point where physiological and behavioral isolating mechanisms prevent the interchange of genes. If individuals from either Zone B or Zone C migrate back to Zone A, they can occupy this zone together with the Zone A species without interbreeding and losing their specific identity.

Fig. 3.1. A model for geographic speciation, thought to be the dominant mode for the evolution of new species.

contrast, stay-at-home naturalists saw very little variation within the local species of animals and plants. Other naturalists observed similar cases where what appeared to be a single species was found on both sides of some barrier such as a desert or a mountain range. In each of these cases the gene pools would be at least semi-isolated, each carrying its allotment of the original species' *gene pool*.

By the 1930s and 1940s the evolutionist Ernst Mayr had made a convincing case for *geographic speciation*: The recently isolated population might find itself in a region where temperature, rainfall, vegetation, predators, or food supply differed from that of the original environment. If some of the immigrants could survive the new conditions, the forces of selection would be different and very strong—as in the case of insects first confronting pesticides. The new population's size would be much smaller than that of the original population, and as hostile environmental conditions were confronted its mortality rate would be greater. With fewer individuals to breed, more inbreeding would occur and recessive traits would have an opportunity for greater expression. In time the newly isolated population might become so unlike the original population as to be recognized as a different species.

In the case of some species, isolation need not involve major ecological or geographical distances. Many species of parasites, for example, are restricted to a single host species, while closely similar species occur in other hosts. For instance, it is suspected that the protozoan genus *Plasmodium*, which causes malaria, was originally a generalized parasite, but in time its populations isolated specifically in birds, human beings, or other animals were each selected for improved survival in a specific host. Different host species provided the equivalent of geographic separation. After all, there are considerable *barriers* to cross to get from the circulatory system of a bird to that of a mammal. The spread to a new host probably occurred in the same manner by which the malaria parasite moves from one human host to another—it is carried by mosquitoes.

Evolutionary biologists are fairly comfortable with the hypothesis that evolution can occur as a consequence of usually great geographic isolation as well as *microgeographic isolation* on different host species. However, there is vigorous discussion but little agreement today about the possibility of other, presumably minor, mechanisms.

Once two populations become genetically different, there are other mechanisms for keeping their gene pools separated. For example, closely related species living in the same region may differ in the times of year that they breed, or they may have anatomical differences that prevent *interbreeding*, or they may have a behavioural preference for breeding with members of their own species or be restricted to different local habitats such as a forest or grassland. There is also the powerful mechanism of genetic incompatibility. In most cases individuals of even closely similar species cannot produce viable offspring when cross-mated. Fertilization may occur, but the embryos may die early in development. Some hybrid crosses go a step further. The embryos develop normally, but they become sterile adults. Female horses and male donkeys, for example, can breed and produce mules—prime examples of hybrid vigor. But mules do not produce normal eggs and sperm. Their set of horse chromosomes and another of donkey chromosomes differ enough from each other so that they cannot participate in the formation of normal sex cells. These various reproductive mechanisms all help to prevent interbreeding between closely related species and preserve the genetic integrity of each.

What is a Species?

Consider the following: In a series of fossils belonging to the same lineage, each subsequent generation is imperceptibly different from the parent generation, but after many generations the individuals may become recognizably different in structure from the original species A. If those differences are similar in magnitude to the differences that separate living species, we would call the later population species B and we would say that species A evolved into species B. There was never a time when both species were present. Still, to be strict about defining species, we would have to arbitrarily draw a line at some generation and say that the offspring of species A are now members of species B. This is never a problem because the gaps in the fossil record eliminate the intermediate forms.

That, of course, is a ridiculous. A consensus on the definition of species might be easier to reach with respect to living species, but even in these cases, various criteria are used to separate species and no single criterion is adequate in all cases. The most useful criteria are those associated with the ability or inability of populations to interbreed, which horses and mules can do, and freely exchange genes, which they cannot do. If two populations in the wild do not exchange genes, each can maintain its *genetic integrity* and evolve in its own way. If they can *exchange genes*, their genetic integrity will be lost. It is assumed that individuals of the same sexually reproducing species belong to a single *interbreeding unit*; that is, any individual can in theory breed with any other individual of the opposite sex and produce viable offspring. That is a good criterion, but it cannot be used when two populations of individuals that might be capable of interbreeding are separated by a *geographical barrier*—such as when one population lives on the mainland and another on an offshore island. The two geographically isolated populations could not cross because there would be no opportunity for boy to meet girl. If the two isolated populations differ sufficiently from one another in ways that usually characterize different species, each could be regarded as a distinct species; if not, they would be thought of as a single species isolated into two groups.

On the other hand, the ability of two populations to interbreed under natural conditions does not mean that they are always recognized as a single species. A notable example is lions and tigers. They can be crossed in captivity, and depending on the direction of the cross, the offspring are *ligers* (male lion and female tiger) or *tiglons* (male tiger and female lion). This sorrowful miscegenation is avoided today because lions are largely restricted to Africa and tigers to Asia. Some species of wild ducks, very different in external characteristics, do interbreed but not to the degree that they cease to be different species. Thus, the individuals of a species are usually part of a single potentially interbreeding unit, usually do not interbreed with individuals of another species, are usually distinguishable on the basis of external characteristics and behaviour patterns, and are usually found in a restricted geographic area where all individuals can, in theory, meet all other individuals of the same interbreeding unit. It is fascinating to contemplate what would have been the nature of biodiversity today had extinction not been such a dominant force in the history of life—that is, if examples of every species of micro-organism, plant, and animal that ever evolved remained to this day. The result would be a complete intergradation of all life, and the recognition of species would be totally arbitrary. Bacteria and elephants would be connected by populations that formed a continuum of nearly identical types. The fossil record shows, however, that every species that ever lived either evolved into another species, evolved into two or more species, or became extinct. Thus life today is not a continuum but is sequestered in distinct packages—the many millions of different species.

Classification

At the time Darwin was working on the Origin in the 1840s and 1850s, known species of plants and animals were arranged in a *hierarchical classification* based almost entirely on their structural characteristics. According to Darwin's theory, the similarity of species within any category of classification was due to their evolution from a common ancestor, and this is the explanation we still use today. But the current classification of living organisms, especially in poorly studied groups, is far from settled.

One basic problem is the lack of sufficient fossil data in most lineages. Another is the uncertainty surrounding selection of characteristics as the basis for classification. Most categories of classification of living species are based on their shared common characteristics and on their differences from species in other groups. Thus, the categories of classification include the likes and exclude the unlikes. This may sound vague, and it is. Classification is an art that depends on the knowledge and skill of the classifier. But the classification of species in well-studied groups, such as the flowering plants, vertebrates, and many kinds of arthropods, has reached a considerable level of stability. This means that competent *systematicists*—professionals whose job is to classify organisms—tend to agree on the basic classifications.

In the next few decades puzzles about classification that have persisted for generations have a good chance of being solved at last, as the new techniques of molecular biology provide quantitative data on the degrees of resemblance between two organisms. One method compares the similarities of DNA in different organisms. The long strands of DNA are extracted and broken into short segments. The DNA of two different kinds of organisms can be mixed. Since similar segments of DNA are able to combine with one another, the proportion that combines can be determined. The percentage combining is taken as an indication of relationship—the greater the percentage, the closer the relationship. When Sibley and Ahlquist (1987) used this technique of DNA-DNA *hybridization* to estimate the closeness of various primates, they found that human beings are much closer to the chimpanzee than to any other primate. Progressively less similar were the gorilla, orangutan, gibbon, and Old World monkeys. Two other nucleic acids, *mitochondrial DNA* and *ribosomal RNA*, are also being used to estimate degrees of relatedness. The procedures are quite different, but they and the DNA-DNA hybridizations usually give similar results—an important affirmation of the probability that both are correct.

The *ribosomal RNA* data are now judged to be superior to data based solely on anatomical features. Anatomical differences are usually qualitative and subjective—the shape of a bone, for example—but molecular data can be strictly quantitative. Using ribosomal RNA molecular geneticists have discovered that the traditional order Insectivora, far from being a natural group, is a heterogeneous collection of species. The ribosomal RNA of the elephant shrew, which in the nineteenth century was assigned to the Insectivora, is very different from that of the common shrew, which is a typical insectivore. It was found to be much closer to the elephant, manatee, and hyrax.

The classification of organisms should soon reach a level of precision never before achieved. The data from molecular studies such as those just described will be integrated with the older data from comparative anatomy and embryology, and the existing system of classification will be adjusted accordingly. Systematics, which has been regarded by some as a backwater of biological research, has suddenly become one of the most productive and exciting fields in biology.

Origin of Life

Life had to originate before it could evolve, but we have only vague hypotheses for how this may have happened. There will probably never be direct evidence for how life began, and even if it should prove possible to synthesize life in a test tube, we would know only one way it can be done today, not how it actually happened billions of years ago. The question is so basic and interesting, however, that attempts have been made to study what the contributing events might have been.

The pattern of *origin-of-life* research has been to identify those special properties that distinguish living creatures from nonliving things and then to search for experimental conditions that will allow nonliving systems to exhibit some of these properties of life. A striking difference between the nonliving environment and living systems is in chemical composition. The nonliving environment—the inorganic world—consists of all of the nearly 100 natural chemical elements. The more common ones—iron, oxygen, silicon, and magnesium—together make up about 92 percent by weight of all elements in the Earth's crust. In nature the elements almost never consist of single atoms but are combined to form molecules, which may be of the same or different elements. The vast majority consist of combinations of a few elements. The oxygen in our atmosphere is a molecule composed of two oxygen atoms—O_2. The water molecule is a compound composed of two hydrogen atoms and one oxygen atom—H_2O.

Living organisms do contain some small molecules. In fact, water, the environment in which almost all chemical reactions take place, is the most abundant molecule in cells. However, the molecules that are unique to living organisms are very large, consisting of hundreds or even thousands of atoms. Only six elements—oxygen, carbon, hydrogen, nitrogen, sulfur, and phosphorus—plus a few other elements combine to form the main classes of organic compounds: proteins, carbohydrates, lipids, and nucleic acids. Compounds containing carbon are responsible for the properties that we call life.

A key feature in the origin of life, therefore, must have been a process, or processes, that converted simple inorganic compounds of the Earth's crust into complex organic compounds. Replicating such a process proved a difficult problem for chemists, and for a long time it was assumed that only living organisms could synthesize organic compounds. Then in 1828 Frederich Wohler synthesized from inorganic chemicals an organic compound. It was urea, which is composed of only eight atoms—NH_2CONH_2.

There is no known situation in the world today, apart from living organisms who do it all the time, where complex organic compounds are being synthesized from *inorganic substances*. In trying to understand how life might have begun, we must search for the conditions under which complex organic compounds could be made by a *nonliving system*. If no such conditions exist today, could they have existed in the past? By the middle of the twentieth century, cosmologists thought that conditions during the interval between 4.8 and 3.8 billion years ago, shortly after the Earth was formed, might have been such that organic compounds formed *spontaneously*. That period has been well-named the Hadean eon because of its resemblance to Hades. It was a violent time of tremendous electrical storms as the molten surface of the Earth cooled to form solid rocks. For a while, showers of meteorites bombarded the Earth.

Could complex organic molecules have been formed spontaneously from simpler molecules under conditions thought to prevail in the Hadean eon? In 1953 Stanley Miller performed a simple experiment to test this hypothesis. He put water (H_2O), hydrogen gas (H_2), methane (CH_4), and ammonia (NH_3) in a flask with tubes so arranged that when the water boiled, the steam containing the other molecules passed through a section of the tubing bombarded with electric sparks. The steam was then cooled to form water and returned to the flask to repeat the process.

After a week of boiling this mixture, the water had changed from colourless to coloured. Something had happened. Analysis showed that *amino acids*, *lactic acid*, and other organic compounds that are the building blocks for more complex organic molecules were present. This critical experiment has been repeated many times, and all of the basic types of molecules found in living cells—amino acids, the smaller molecules from which DNA and RNA are made, and sugars—have been produced under the conditions thought to have existed on the early Earth.

In other similar experiments a higher level of complexity was reached when amino acids linked to form small proteins. Quite recently it has been discovered that RNA molecules can replicate themselves outside of a living cell. Not only that, RNA can stimulate the joining of amino acids to form *proteins*. These properties make RNA a candidate for being one of the first *self-replicating* molecules and, hence, a critical step in the *origin of life*.

The working hypothesis of those studying the origins of life is that the sorts of conditions that prevailed in Hadean times—and simulated in *Miller's flasks*—led to the oceans comprising a rich aggregation of organic compounds, a primeval soup. An important step was still required, however. Life is not a stew of oceanic dimensions; it exists in packages. The basic package is a cell. A cell consists of an integrated and interacting set of organic compounds in water, surrounded by a membrane. The membrane keeps the contents intact and prevents many external substances from entering the cell while allowing the needed molecules to enter and waste molecules to exit. The basic composition of the cell membrane is much the same in all living cells: they are composed of proteins and fatlike molecules known as *phospholipids*. In some very recent experiments phospholipids have been synthesized under simulated early Earth conditions, and these are able to join one another to form tiny vesicles. Furthermore, these lipid vesicles can spontaneously take in complex organic molecules, including proteins and DNA, from a surrounding solution.

These and many other experiments are exciting to those who seek to show how life could have originated. It is important to remember that these experiments do not show what happened at the close of the Hadean eon, but they do show that in laboratories today some basic steps in the origin of life can be simulated and studied. Even if a few decades from now it is possible to produce in the laboratory an artificial "*cell*" that can reproduce, this again will show only what might have happened. However, such a triumph would be satisfying to many as probably the best that can be achieved.

The implication so far is that the life of planet Earth originated on Earth. Possibly yes, possibly no. In recent years some meteorites have been found to contain organic compounds that could have been synthesized under purely physical and chemical conditions but also could be associated with life of some sort. These meteorites are thought to date to the time our solar system was forming and to be the detritus left over after the sun and planets were formed. One can even speculate that the origin of life could have involved similar steps whether on the Earth or elsewhere. The astonishing thing is that we seem to be on the threshold of developing the technology and science to answer this question.

Origin of Multicellular Animals

The appearance of animals and plants consisting of many cells was one of the most important moments in the history of life. This event, which paleontologists now believe occurred about 1.7 billion years ago, initiated the extraordinary evolutionary radiation that populated the land and the seas with the creatures familiar to us. Before this, all life had been in the single-cell stage for the first three-quarters of its existence.

Little is known about the origin of the multicellular animals, called *metazoans*, and essentially all of the information has been obtained since World War II. Recently metazoans have been found in the *Precambrian* geological strata, but as yet they offer little information about the origin and evolutionary diversification of the animal phyla.

The first major discoveries of these Precambrian metazoans were made in the *Ediacaran Hills* in South Australia in the 1940s by the Australian geologist R. C. Sprigg. The strata date from about 670 million years ago—approximately 100 million years before the onset of the *Cambrian*. The fossils are mainly impressions of dying animals left in the mud at the bottom of a large body of water. Some paleontologists believe they are primitive jellyfish. Other remains are interpreted as the burrows of wormlike marine organisms. Subsequent to the discoveries in Australia, *Ediacaran*-like fossils have been found in other parts of the world. The Ediacaran organisms are so different from those appearing in the *Cambrian period* and later that paleontologists have yet to agree on how they should be classified. Following the period when the Ediacarans lived, the Earth passed through a time of intense cold, and this may have exterminated most metazoans then alive. The survivors were the ancestors of the species whose hard parts are found as fossils in the *Cambrian strata*.

A noteworthy discovery in 1909 of metazoan fossils of the middle *Cambrian period* was made in the Burgess Shale of Western Canada by the American paleontologist Charles D. Walcott. Many of the fossils are of soft-bodied animals, but a few have shells. Preservation of soft-bodied animals is rare at any period, so it is of great significance to have a fine sample from the time when the major animal phyla are thought to have first appeared and begun their diversification. In recent years the site and the specimens have been intensively restudied, and about 170 species are currently recognized. In contrast with the puzzles of the Ediacaran species, many of the Burgess Shale creatures can be identified as members of the major phyla that persisted and are with us today, such as the sponges, coelenterates, mollusks, annelid worms, arthropods, echinoderms, representatives of other small invertebrate phyla, and even one chordate. Primitive plants were also present. Explaining the sudden appearance of these advanced animal phyla, given their absence among the *Ediacaran fauna*, is a major challenge for future research.

ORIGIN OF HUMAN BEINGS

Darwin knew of no acceptable evidence for the antiquity of the *human species*. Yet it was hard to exclude human beings from the operations of evolution. Darwin almost ignored the question in the first edition of the Origin (1859), knowing full well how difficult it would be for such a notion to be accepted. It was not until near the end of the last chapter that he mentioned the unmentionable. In discussing how the concept of evolution could provide a rational explanation for so many of nature's puzzles, he wrote, "Light will be thrown on the origin of man and his history". Many of those who understood what Darwin was implying might not have eagerly awaited the shedding of that light.

Interest in human antiquity started long before Darwin. By the middle of the nineteenth century a huge array of fossil mammals had been discovered, yet no remains of ancient human beings had been recognized as such. In fact, many doubted that any would ever be found. The French naturalist Cuvier, renowned for his knowledge of fossils, had proclaimed that there were no fossil human beings. This conclusion, perfectly valid at that time, was based on the absence of human bones in association with the bones of extinct fossil mammals. His verdict was accepted by those who adhered to the Judeo-Christian tradition. Yet there were hints that human beings may have existed. For a long time throughout Western Europe people had been aware of the presence of small hammerlike stones that seemed to have been deliberately chipped, and it was hard to believe that other than human beings could have done so. One suggestion was that they were petrified lightning bolts. Others thought they might be tools, but there was no evidence, such as fossil bones, at that time for the existence of human beings who could have made such crude tools. The evidence then available suggested that copper, bronze, and iron were the materials from which tools had always been made. The answer to this problem came when the first European explorers reached the New World in the fifteenth century. They found that the inhabitants made many of their tools from stone, not from metal. Who, then, were those early *Stone Age Europeans*?

In 1856, however, some strange-looking human bones were discovered under about five feet of mud at a site in the *Neanderthal* (valley) in Germany. They were studied and described by a skilled human anatomist, Hermann Schaaffhausen, who suggested that they were of human origin. He noted the thickness of the bones and the shape of the top of the skull, which was somewhat apelike, and suggested that the bones were of one of the ancient savage tribes that had lived in Germany. Others suggested that the bones were not old but merely deformed by some pathology. Another hypothesis was that they were the remains of a soldier in the Russian army that had marched through the area in 1814 on the way to Waterloo.

This was the first good evidence of the remains of an ancient human, now known as "*Neanderthal man.*" Many other individual bones or parts of skeletons were found later, as were the remains of other early European human beings, the *Cro-Magnons*. Although both the *Neanderthals* and the *Cro-Magnons* seemed to be more ancient than 4004 b.c., their ages could not then be surmised. The *Cro Magnons* were structurally like modern human beings, and the *Neanderthals* were only slightly less so.

It was not until the closing years of the nineteenth century that really old human remains were discovered. In 1891 a Dutch physician, Eugene Dubois, began an active search in Java for fossil remains of early human beings. A crew of convicts was assigned to help with the digging. At a depth of 40 feet they found a human tooth and the top of a skull; a year later, they uncovered a leg bone (femur). The skull was so primitive that at first Dubois identified it as a fossil chimpanzee. But the shape of the femur indicated a creature that walked upright—therefore more likely human than ape. Dubois estimated that the capacity of the cranial cavity was about 900 milliliters, compared with 1,300 milliliters for modern human beings and 400 milliliters for the chimpanzees. The fossil was given the name *Pithecanthropus erectus* ("*ape man erect*") to indicate its intermediacy between apes and human beings and that it walked on two feet. The name was later changed to *Homo erectus*.

This was the first discovery of a truly ancient human fossil. On the basis of later information *Homo erectus* was estimated to have lived 1.7 million to 250,000 years ago. It was assumed that Homo erectus evolved from a much more ancient ancestor that was also the progenitor of the modern great apes, the gorilla, orang, chimpanzee, and the bonobo. By the close of the nineteenth century most paleoanthropologists agreed that the *Homo erectus*, the *Neanderthal man*, and modern human beings form a sequence on the path of human evolution. But as the twentieth century was to reveal, the story is more complicated than that.

The fossils of *Homo erectus* and the *Neanderthals* established that individuals intermediate in structure between some unknown prehistoric apelike creatures and modern human beings had existed—as Darwinian theory demanded. This indicated that humans had evolved, but not that those two fossil types were direct ancestors of present-day human beings. Far more material would be required to establish such relationships. The search for evidence of our past history became more intense, but for several decades the discoveries were few—though more *Neanderthal* and *Cro-Magnon* remains, as well as a single jaw of another species, *Homo heidelbergensis*, were discovered in Europe.

In the 1920s important discoveries began to be made in Africa. The first, in South Africa, was of an immature individual with features intermediate between those of apes and *Homo*. It was given the name *Australopithecus* ("*southern ape*"). This was the first of an extraordinary group of African fossils that has thrown considerable light on our past history.

Possibly half a dozen well-defined species in the genus *Australopithecus* are now known, all from Africa. *Australopithecus* was a way station in the evolution from the very ancient apes to species of the genus *Homo*. The size of the brain in the *Australopithecus* species was 500 milliliters or less, only slightly larger than that of a chimpanzee.

One of the species of *Australopithecus* was probably ancestral to Homo habilis, generally regarded as the first species of Homo to evolve. This event occurred about 2.4 million years ago. Existing data suggest that Africa remained the hub of human evolution for thousands of years, but gradually various species migrated to Asia and Europe. Even *Homo sapiens* is thought to be African in origin, and it is the species that includes all modern human beings.

The data now show beyond all reasonable doubt that human beings have evolved over millions of years from some ancient primate that was more apelike in skeletal characteristics than are modern human beings. This means that since about 6 million years ago when the lineages leading to human beings and to the great apes diverged, ours evolved much more rapidly. Yet by some measures we are not all that different from other advanced primates. We share about 98.4 percent of our genes with the chimpanzee. We might say that the chimp is 98 percent human, or that humans are 98 percent chimp. In fact, our degree of similarity to the chimps in biological characteristics has led a prominent biologist, Jared Diamond (1992), to suggest that human beings are really a third species of chimpanzee, joining the common chimpanzee and the pygmy chimpanzee. Thus we would abolish the genus *Homo*. Nonbiological reasons will probably prevent Diamond's suggestion from being widely adopted. Alternatively, we might point out that a mere 2 percent difference in genetic makeup can have astonishing consequences.

Human evolution is a complex and rapidly advancing field of research and will remain so as long as only the estimated 3 percent of all extinct primate species have been discovered. Although information about human evolution is increasing rapidly, the data are not yet sufficient to be sure of the exact lineages. It is wiser to accept the usual caution of paleontologists and speak of intermediate or transitional forms rather than *ancestors* and *descendants*. The fossils so far uncovered form a series of transitional anatomical types, with the more ape-like ones being the oldest and those more like Homo sapiens being the most recent. There is another problem, so far without any generally accepted answer, and that is to understand the origin of the different types of human beings, all considered to be a single species, *Homo sapiens*, now occupying the Earth.

Thus, in spite of the important discoveries, especially in the last half of the twentieth century, a great deal still waits to be learned about evolution in general and our own in particular. But this, of course, is typical of all science: a problem solved gives us an answer but leaves us with new questions that extend and refine the target of our inquiry.

4

Genetic Variations

The type of symmetry displayed when a sentence or design is repeated is known as *translational symmetry*. Remember that the symmetry of an object can be defined by the number of different transformations that can be applied to it that leave it unchanged. We have already come across *rotational symmetries*, in which an object is left unchanged when it is turned; and *reflection symmetries*, in which appearances are preserved following reflection in a plane. An object with *translational symmetry* doesn't change when you move or shift it along in a straight line by a certain amount. Imagine, for example, that the design carried on for an infinite distance to the left and right. If you were to shift the pattern along by one length of the repeating unit, you would end up with exactly the same pattern again. In other words, you would not be able to tell that such a transformation, called a *translation*, had been applied by simply looking at the outcome. Similarly, an infinitely long page of written '*lines*' would not be altered by shifting it up or down by one line. Strictly speaking, only infinite patterns show this sort of symmetry. If the repeating pattern is of a limited length you could tell that it had been shifted along by observing the movement of its ends. Nevertheless, for our purposes it will be useful to apply translational symmetry in a looser sense to include finite as well as infinite repeating patterns. After all, you can always imagine converting a finite pattern into an infinite one by extending it indefinitely.

Translational Symmetry

With this notion of *translational symmetry* in mind, we can distinguish between two basic types of repeating pattern. As well as

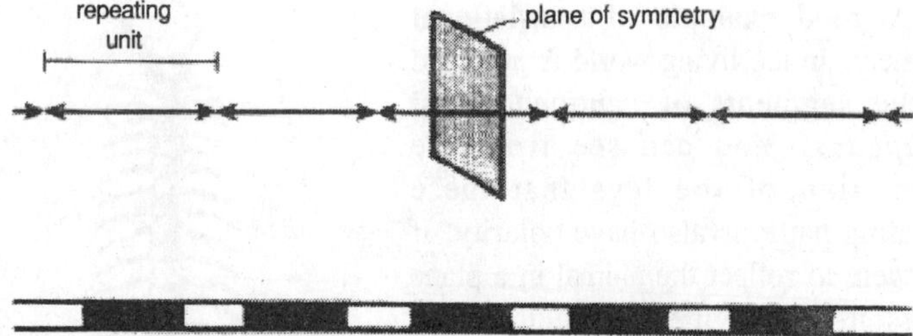

Fig. 4.1. Translational symmetry with no polarity.

showing translational symmetry along the horizontal axis, the pattern has a vertical plane of reflection symmetry running down the middle. That is to say, the left and right halves of the design are mirror images of each other. If the pattern was repeated indefinitely, there would be a similar plane at regular intervals all the way along (the interval between the planes of symmetry would be half the length of the repeating unit). Such patterns can be said to show no polarity because they look the same either way round. A band of two alternating colours also has this sort of symmetry, with the mirror planes running down the middle of each colour block.

The second type of repeating pattern has no such planes of symmetry. For example, if the pattern is reflected in any vertical plane, the direction of the arrows will be reversed, allowing you to tell that the transformation has been applied. It therefore has a lower degree of symmetry than the previous examples because there are fewer ways of transforming it into itself. Such patterns can be said to exhibit polarity. A banding pattern of three different colours has this sort of symmetry because any reflection at right angles would invert the order of the colours. All cases of *translational symmetry* fall into one or other category, either having polarity or not. A stack of similar coins oriented in the same way, say heads up and tails down, would have translational symmetry with polarity; each coin would be related to the next by a vertical shift of one coin's thickness, and the direction each coin was facing would be inverted by reflection in any plane at right angles to this (a horizontal plane in this case). By contrast, a stack of double-headed coins (coins with heads on both sides), would form a pattern showing translational symmetry without polarity.

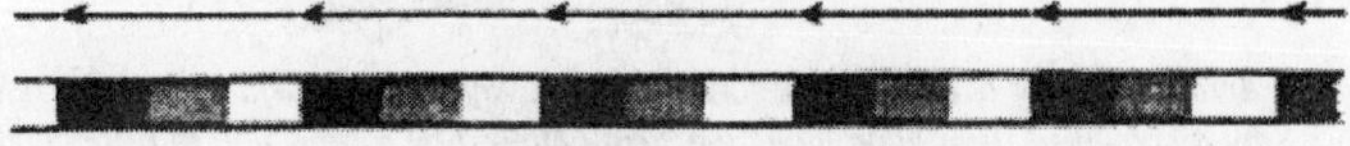

Fig. 4.2. Translational symmetry with polarity.

A good example of translational symmetry in the living world is provided by the segments of *centipedes* and *millipedes*. You can see from the orientation of the legs that these repeating patterns also have polarity: if you were to reflect the animal in a plane across it, the segments would end up pointing the opposite way. Leaves on most plants are also arranged in a repetitive way, although they usually display a more complicated pattern than pure translational symmetry. In some plants the leaves are arranged as a spiral or helix up the stem. To get from one leaf to the next you need to shift up the stem (*translation*) and move part way round its *circumference* (*rotation*). That is to say, the pattern remains unchanged following a combined translation and rotation. In other cases, as with the begonia, subsequent leaves are related by a translation up the stem followed by reflection in a longitudinal plane. Most types of leaf arrangement involve

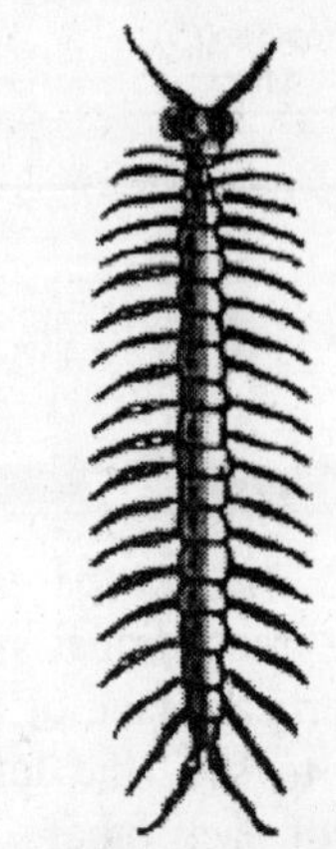

Fig. 4.3. Centipede showing translational symmetry of segments.

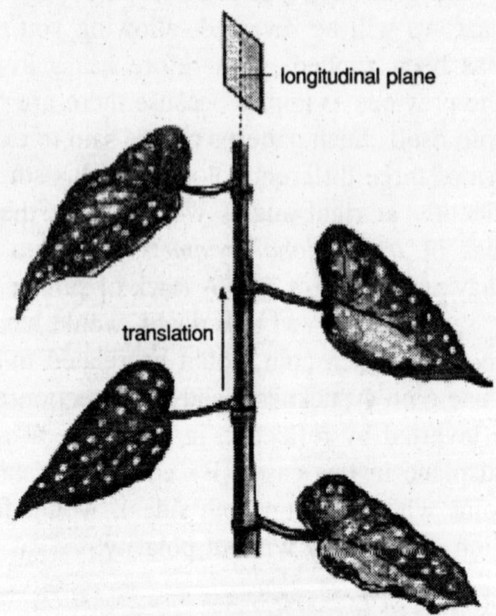

Fig. 4.4. Arrangement of leaves on a begonia plant, in which subsequent leaves are related by a combined translation along the stem followed by a reflection in a longitudinal plane

a combination of translation, rotation and/or *reflection symmetry* (they also show polarity because the upper surface of a leaf is usually quite different from the lower surface).

All these examples involve repetition of the same or very similar units. But there is also a broader category of patterns in which repeating units vary to some extent from one to another. Suppose, for example, that instead of identical coins we stack a mixed set of coins of different denominations. The coins can still be stacked with the same polarity (heads facing up and tails down) but there is now no translational symmetry because each coin is distinct from its neighbour. Furthermore, each coin may have a slightly different thickness so that there is no standard length for shifting them up or down. Nevertheless, all the units in the stack share some features because they are all easily recognizable as coins. One way of describing the situation would be to say that there is a basic theme that is repeated up the stack— the general notion of a coin—but there are also important variations on this—the particular denominations.

The same point can be illustrated with a queue of people. If all the people in a queue are identical and stand at perfectly repeated intervals, the queue would show translational symmetry. But in real queues, each person is different from the next, so there is no strict symmetry of this type. Nevertheless, we can still recognize a repetitive structure in the queue, which might be expressed by saying there is a basic theme—the general notion of a person—that is repeated along the queue, together with numerous variations upon it—the particular characteristics of each individual in the queue. In all these cases a basic type of unit, such as a coin or person, is repeated in a linear sequence but the repeating units are not identical.

A fruit fly can also be thought of as exhibiting variations on a theme. We talk about a fruit fly as being made up of repeating segments even though they are clearly not all the same: some have legs, some also have wings, yet others have no appendages. It is as if the fly displays a repeated theme, the segment, combined with variations upon it. The same might be said about the different whorls of organs in a flower, or the fore and hind limbs of a vertebrate: they are all examples of variations on a theme.

Explicit and Implicit Themes

In 1957, Pablo Picasso produced a series of paintings inspired by one of the masterpieces of art, Las Meninas (The Maids of Honour), painted about three hundred years earlier by Diego Velazquez.

Velazquez's painting shows a girl of the Spanish royal family, attended by her maids of honour. The girl has come to watch the Court painter, Velazquez himself, at work in his studio. Picasso was fascinated by the picture and produced a series of paintings based upon it. Some of Picasso's paintings are based on the whole of Velazquez's picture; others concentrate on a detail such as the girl alone. Once you know the history of the paintings, you can almost see the original Velazquez within them, even though on the surface they look so different.

This series of paintings by Picasso might be said to represent variations on a theme. The theme is the original painting by Velazquez whereas the variations are the different versions of it that Picasso came up with. Although the theme and its variations are all paintings, there is no problem in distinguishing between them. The theme dearly preceded the variations and was their source of inspiration rather than the other way round.

But there is also another theme that unites Picasso's paintings: they have a common style. They are all examples of Picasso's art and vision. In a sense this is a more important theme because anyone can look at a Velazquez and produce a series of paintings, but only Picasso could have produced the works that we see in his Las Meninas series. Yet defining this theme, Picasso's style, is not so easy. You might look at many examples of paintings by Picasso and recognize some unity, but it would be difficult to say precisely what they all have in common, what makes them all identifiable as works of Picasso.

Picasso's Las Meninas series illustrates two basic types of theme. On the one hand, there is the theme provided by the Velazquez. We can point to Velazquez's original painting as distinct from the variations based upon it. We might say that in this case the theme is quite explicit and precedes the variations. By contrast, the other type of theme, the stylistic unity of Picasso, is implicit. It seems to be locked within each of the variations rather than being a separate painting we can point to. Unlike an explicit theme, there is no implied direction here: any two Picasso paintings are equivalent expressions of Picasso's style, rather than one being derived from the other.

Leaf

Much confusion about the nature of repetition in the living world has come from confounding implicit with explicit themes. A good illustration of this concerns Goethe's idea that all the appendages of a plant are equivalent to leaves. Recall that he proposed that all the organs of a plant were variations on a theme: sepals, petals, stamens, carpels and foliage leaves may look different from each other but they share an underlying similarity. Goethe wanted a term to describe the theme that unified the various organs of a plant and he chose to use the word leaf:

> It goes without saying that we must have a general term to indicate this variously metamorphosed organ, and to use in comparing the manifestations of its form; we have hence adopted the word leaf. But when we use this term, it must be with the reservation that we accustom ourselves to relate the phenomena to one another in both directions. For we can just as well say that a stamen is a contracted petal, as we can say of a petal that it is a stamen in a State of expansion. And we can just as well say that a sepal is a contracted stem-leaf ... as that a stem-leaf is a sepal.

Goethe was using the word *leaf* in an unusual way, to refer to the implicit theme behind all organs. There was no implied direction: a comparison between any pair of organs, like a stamen and a petal, or a sepal and a foliage leaf, could always be made both ways. This equivalence could be summarized by saying that they are variations on the same theme, all different types of leaf. Unfortunately, in choosing the word leaf to denote the underlying theme, he caused major confusion because we also use leaf to indicate one of the variations, namely foliage leaves. By saying that all organs were essentially leaves, he did not mean to imply that leaves, in the narrow sense of green foliage leaves, were an explicit theme upon which everything was based. He was using leaf in a more implicit sense, as a theme common to foliage leaves and floral organs. His notion of leaf was more like the stylistic unity of Picasso's Las Meninas paintings than the distinct Velazquez that preceded and inspired them.

Even though he was very careful to point this out, it did not stop people misinterpreting his theory. Many botanists believed Goethe to be saying that the different parts of a flower were derived from leaves in the common sense of foliage (this mistake is still sometimes made even today). The green leaf of a flowering plant was taken to be the template that everything else was based upon. After Goethe died, this same misinterpretation surfaced again but in a slightly different way, under the guise of evolution. As people became convinced of Darwin's theory of evolution in the late nineteenth century, they started to consider what the ancestors of flowering plants might have looked like. If the floral organs were indeed modified foliage leaves, as they thought Goethe was saying, it would seem that leaves came first and flower parts evolved from them secondarily. The implication was that ancestral plants must have had only foliage, and flowers then evolved through the modification of this ancient greenery.

The trouble is that if these ancestors had only had foliage leaves, they would have been completely sterile. A rose leaf cannot produce sperm or egg cells, no matter how long you wait. Nobody was around to witness the ancestors of flowering plants but it is extremely unlikely that they were sexually sterile. *Sex* is one of the oldest and most ubiquitous inventions of evolution and greatly predates the arrival of flowering plants. It is therefore highly improbable that the early ancestors of flowering plants had only sterile organs. They are more likely to have been similar to modern *fern* plants. Ferns do not have flowers but they do reproduce sexually. They achieve this through the production of sexual spores from their green fronds or '*leaves*'. These spores will eventually give rise to sperm and egg cells which can fuse to produce the next generation of fern plants. In other words, the fronds on a fern resemble both foliage leaves, being green and spread out to collect the energy from the sun; and flowers, being the route to sexual reproduction. In flowering plants there is a separation between organs involved in nourishment (*foliage*) and sex (*flowers*). It seems likely that this is a later specialization, and that both functions were carried out by the single type of organ in their early ancestors.

Some botanists therefore concluded that Goethe must have been wrong when he said that the parts of the flower were modified or metamorphosed leaves. It seemed to imply that ancestral plants were just sterile foliage, which was obviously incorrect. As the botanist F. O. Bower said in 1911: 'There is no need now for any theory of metamorphosis to explain the origin of the Flower. As first stated by Goethe it was a theory of mysticism, which was doomed to dissolution.' But this is because Goethe's notion of leaf was taken to be an explicit theme from which other organs were derived. Goethe was actually using leaf to denote an implicit theme that was simply manifested in different ways in the various parts of the plant. According to this viewpoint, the leaves of ancestral plants would have represented yet other variations on the same theme.

Even when taken on his own terms, though, Goethe's view seems to lead to an impasse. What exactly is a leaf, this common theme that is continually being depicted by the plant? There seems to be no meaningful way to pursue the problem other than by saying it is that which is common to all plant organs. This just seems to be evading the issue. It was for such reasons that Goethe's ideas were often considered to be mystical idealism rather than serious science.

Is there any way of getting beyond this impasse? Although the stylistic unity of Picasso's paintings is difficult to describe, we know where its origin lies: with Picasso himself. The implicit theme of his paintings reflects the way Picasso's brain worked and interacted with the canvas as each painting was produced. This means that one way of getting a deeper insight into the implicit style of his work might be to try and understand more about how Picasso's paintings arose, what led Picasso to apply the paint in one way rather than another. In a similar way, to define the implicit theme that is common to petals, stamens and foliage leaves, it is no good looking at more and more examples of fully developed plant appendages. We need to look deeper, for the common internal processes that lie behind the development of these organs.

Repeating Colours

If you look at the embryo of a fruit fly at a relatively late stage of its development, certain features can be seen to be repeated along the length of the body. Most obvious is the pattern of thorny outgrowths or denticles. The denticles are most abundant towards the head end of each segment. The segments display a consistent repeating pattern, with a thorny head region and a smooth tail region, as you travel along the length of the body, like a stack of coins oriented in the same way. Ignoring the variation between segments for the time being, the pattern shows translational symmetry because you can shift it into itself by moving it along by one segment. The pattern also has polarity because of the consistent orientation of the head-tail differences of each segment: if the embryo looked at itself in the mirror, another embryo looking in the opposite direction with the polarity of each segment flipped around to point the other way.

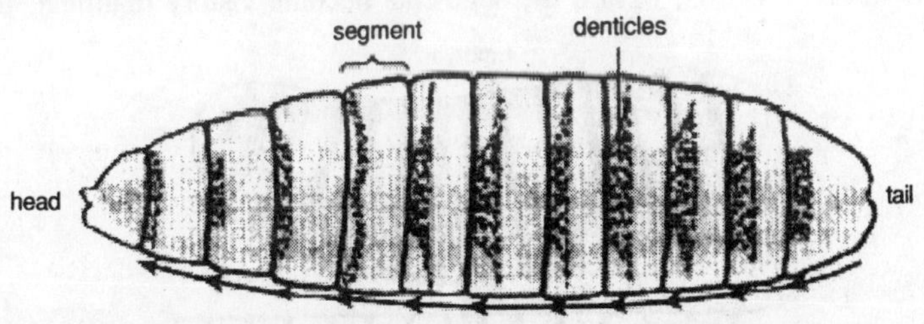

Fig. 4.5. Fruit fly embryo at late stage of development, viewed from the belly, showing repeating pattern of denticles on the segments.

A good way of trying to understand biological patterns is to study mutants in which they are altered. In this case, we would be looking for mutants with alterations in some aspect of their translational symmetry. In the late 1970s, Eric Wieschaus and Christiane Nüsslein-Volhard were looking through thousands of fruit fly embryos, trying to find exceptional cases that deviated from the norm in interesting ways. As they went through these screens, they noticed a striking class of mutants. These had the normal number of segments but instead of having a head-tail polarity, each of the segments was symmetrical about its middle (the segments were also shorter). For instance, the thorny outgrowths, typical of the head region of each segment, might be repeated as a mirror image, replacing the tail region, so that the embryo became equivalent to a stack of double-headed coins. It was as if the polarity of the pattern had been lost; Nüsslein-Volhard and Wieschaus therefore named these segment polarity mutants.

It is important to bear in mind that these mutants do not affect the variation between the segments: segment 4 can still be distinguished from segment 5 even though each segment is now made up of two symmetrical halves. In a similar way, you could imagine replacing a stack of coins with different denominations by a stack of double-headed coins, while leaving the denominations unchanged. Segment polarity mutants affect the theme but not the variations.

Many of the segment *polarity genes* are expressed (switched on to make RNA copies) in a repeating pattern in the early embryo, corresponding to the pattern of segments that will form later on. The basic repeating pattern is illustrated below the embryo. The region depicted by each of these stripes will eventually form part of a segment, say part of its tail end. Another segment polarity gene might have an expression pattern that is shifted along slightly, such that each stripe corresponds to a different region of each segment, say part of its head end. In other words, before the segments become visibly manifest, the

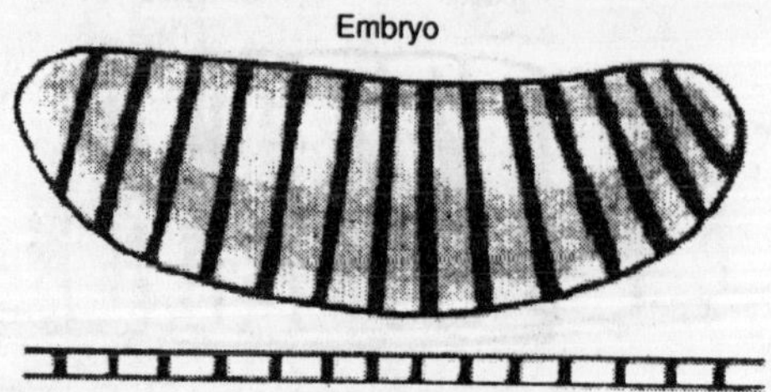

Fig. 4.6. Expression pattern of a segment polarity gene in an early fruit fly embryo, with the basic repeating pattern shown below.

embryo already contains a series of gene activities that divide it up in a corresponding repetitive pattern. This pattern is normally hidden from view but can be revealed by looking at the expression pattern of the segment polarity genes.

As with many of the other genes we have come across that affect *fundamental symmetries*, many of the segment polarity genes code for hidden colours (*master proteins*), associated scents (*signalling proteins*) or sensitivities (*receptor proteins*). We can think of the normal role of these genes as establishing a repeating pattern in the early embryo.

Imagine the early embryo contains repeating stripes of three different coloured regions. One set of three colours makes a repeating unit, corresponding to the length of one segment. To make things easier to remember, colours with the same initial letter will be used as the corresponding region of each repeating unit: *hazel* (head region), *mauve* (middle region), and *tangerine* (tail region). These hidden colours provide a basic frame of reference that is interpreted and elaborated, eventually becoming manifest in the anatomy of each segment. As we saw earlier in this chapter, such a pattern of three colours has translational symmetry with polarity: it can be moved into itself by shifting it along by a repeating unit, but has no planes of symmetry at right angles to this.

Now suppose we have a mutant in which one of these hidden colours, say tangerine, is lost. We end up with a repeating pattern of only two colours, hazel and mauve. As we saw earlier, a pattern of only two alternating colours still has translational symmetry but lacks polarity: a plane of reflection symmetry passes through the middle of each region of colour. By removing one hidden colour, *tangerine*, we have increased the degree of symmetry because there are now more ways of transforming the pattern into itself. The same symmetry will also become evident in the anatomy of the segments that develop by interpretation of this pattern. This is what happens in the mutants with double-headed segments: they lack a key element in the basic pattern of hidden colours and so develop with the corresponding symmetry.

What would happen if a second hidden colour was lost? You might expect such a mutant, with a single colour, to give rise to an animal with uniform segments. However, the change is even more fundamental than this. This is because the anatomy of the animal, including the production of visibly distinct segments, depends on the interpretation and elaboration of hidden colours. By eliminating the repetitive pattern of hidden colours we would also get rid of any anatomical repetitions, such as the development of a visible segment boundary at regular intervals. In other words, there would no longer be any segments to speak of because the boundaries between segments are themselves a later manifestation of the repetitive pattern of hidden colours. This is essentially what happens in mutants in which all of the segment polarity genes are inactive: the embryos lack dear repetitive structure and remain unsegmented.

In mutants with only one hidden colour all the way along, it may look as if there is no repetition left. Strictly speaking, though, we have ended up with a higher degree of *translational symmetry*. A uniform line can be transformed into itself by shifting it along by any amount (assuming it is infinitely long). It is also unchanged by a reflection in any plane at right angles to it. The loss of hidden colours from our original pattern therefore leads to an increase in symmetry: loss of one colour leads to more *reflection symmetry* (lack of polarity), loss of two colours to an increase in both translational and reflection symmetry. Put in another way, the hidden colours increase the degree of asymmetry, much as we have seen in previous chapters.

The segment polarity genes therefore provide the basic theme behind a segment. It is their repetitive pattern that we recognize in a modified and elaborated form in the visible anatomy of the animal. Without these genes, the theme disappears and we are left with a monotonic organism.

Combining Themes with Variations

The repeating theme of hidden colours that are common to segments (hazel-mauve-tangerine). But we have seen in previous chapters that variation between segments also depends on a set of hidden colours: the various types of green that distinguish segments from the head to the tail of the embryo. To see how these two aspects, themes and variations, interact with each other, we need to combine their respective colours. Where the two patterns of hidden colour are simply superimposed to give an overall patchwork. By combining the colours in this way, each segment gets a distinctive overall type of green, while at the same time there is a common pattern to all segments sct by the repeating sequence of hazel-mauve-tangerine. Mutants lacking the green colours have segments, but all of them are identical (i.e. they have a theme without variation). Conversely, mutant embryos that lack the repeating colours do not have segments, but still show variation from head to tail (i.e. they have variation without a theme).

The notion of themes and variations seems much more straightforward when understood in terms of overlapping hidden colours than when we look at an adult fly. This is because the adult is a much later manifestation of this underlying pattern. The patchwork provides a frame of reference that is interpreted and elaborated, eventually becoming manifest in the anatomical features of the fly. However, during this process, the simplicity of the underlying pattern can be extensively transformed so that we perceive it much more dimly and indirectly in the final anatomy. In some cases the transformations can be so extensive that the segments become hardly recognizable at all: the head of a fly seems very different from the rest of the animal, yet much of it is derived from three of these repeating units. The themes and variations we see in the adult are a highly distorted view of a much earlier simplicity.

From Rainbows to Zebras

For this system to work, a theme and its variations need to be tightly coordinated. The two overlapping sets of hidden colour have to be interwoven in a precise way so that the variations and theme can be correctly superimposed. How is this precise coordination achieved?

It has already described how the various types of green are established. Recall that the green colours themselves depend on an earlier pattern of rainbow colours that divide the embryo from head to tail into broad regions of *red*, *orange*, *yellow*, *green*, *blue*, *indigo* and *violet*. The rainbow gives a progressive change from head to tail, and this information is interpreted to give various types of green from one end to the other (apple-green, bottle-green, cyprus-green, etc.). This seems relatively straightforward because we are using one pattern that progresses from head to tail, the rainbow, to generate another pattern that progresses in a similar way, the various greens.

What about the repeating pattern of hazel-mauve-tangerine; where does that come from? This also depends on interpreting the rainbow pattern of hidden colours, but in a rather different way. To illustrate how a progressive rainbow can be used to give a repeating pattern, It will give a picture that is rather simplified but that nevertheless explains many of the basic principles involved.

Like a true rainbow, the hidden colours of the early embryo are not separated by sharp boundaries, but gradually merge from one to the other. The hidden colours overlap with each other, so that where yellow overlaps with green, for example, you get a combined yellow-green colour, because both the yellow and green master proteins are there together. This means that in addition to seven regions of what might be called pure colour, there are also six regions of mixed colour, where the colours from adjacent regions overlap.

Recall that each gene is divided into two parts: a *regulatory region* containing sites where master proteins can bind, and a *coding region* carrying the information used to make a protein. The way a gene interprets a pattern of hidden colours depends on the combination of binding sites in its regulatory region. This will determine where the gene is switched on or off.

To make a repeating pattern, we will need an interpreting gene with binding sites for the rainbow colours. Recall that a binding site is named according to the hidden colour that recognizes it: red master protein binds to an *R-site*, orange to an *O-site*, yellow to a *Y-site*, etc. *G-site*, *B-site*, *I-site* and *V-site*. We also need a rule for how the master proteins affect the activity of the gene. In this case, the rule will be that the gene only gets switched on if two master proteins are bound at the same time (recall that such rules depend on how these master proteins interact with each other and with other proteins bound to the regulatory region). This means that the gene will only be active in regions where the colours overlap with each other: the gene would not come on, for example, in the pure yellow or pure green regions because they only contain one master protein, but it would come on in the yellow-green region because it would have both the yellow and green proteins bound to it. The overall result would be that this gene would come on at every one of the six regions where the colours overlap. (A key piece of experimental evidence in support of this type of model is that if many of the binding sites are deleted, to give a regulatory region with a more limited number of sites—say only a *Y-site* and a *G-site*—the interpreting gene only gets switched on in one stripe, where yellow and green overlap.)

Now suppose that this interpreting gene itself produces a hidden colour, say ebony, from its coding region. This means that ebony will be produced in all the regions where the rainbow colours overlap: we will end up with a repeating pattern of ebony stripes at regular intervals along the embryo. The progressive rainbow pattern has been interpreted to give a repetitive pattern of hidden colour. In a similar way, we may introduce a second gene that codes for a different hidden colour, say ivory. Like the gene for ebony, it also has binding sites in its *regulatory region* for all the rainbow proteins. But in this case, the rule is different: the gene for ivory only gets switched on when only

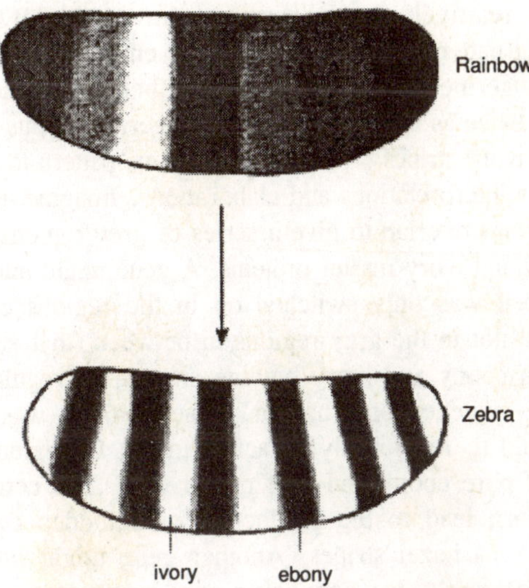

Fig. 4.7. Interpretation of rainbow to give a zebra pattern of alternating stripes in the embryo.

one master protein is bound. If more that one protein is bound, the gene for ivory gets switched off rather than on. This gene will therefore only be switched on in the seven regions of pure colour, and will be off in the regions of mixed colour. It will paint a pattern of ivory stripes alternating with ebony stripes. The rainbow has been interpreted to give a zebra.

The genes for *ebony* and *ivory* correspond to another group of genes that affect early embryo development in fruit flies, *pair-rule genes* (so called because they affect alternate segments along the embryo). The discovery of these genes was a major breakthrough because they provided a way to link the progressive pattern of the rainbow with a repeating pattern. When these genes were isolated, the zebra pattern of alternating stripes could be revealed by staining the embryo (there are actually seven stripes of both ebony and ivory rather than the six stripes of ebony in my simplified model). An unscheduled speaker, Ernst Hafen (working in Walter Gehring's lab in Basel, Switzerland), jumped to the stage and showed a slide of an elliptical embryo with 7 beautiful zebra-like stripes of gene activity running down it, glowing in the dark. There was an audible gasp in the audience. For the first time, we could see that what seemed like a featureless embryo, showing no physical signs of repetition, was actually subdivided into regular repeating regions of gene activity.

It is relatively straightforward (in principle) to get from the alternating pattern of ebony-ivory in the embryo to the stripes of hazel-mauve-tangerine. The main difference between these patterns is that there are twice as many hazel-mauve-tangerine repeats (14) as individual ebony or ivory stripes (7). Getting from one pattern to the other depends on further interpretations and elaborations. Imagine that the ebony and ivory regions overlap to give a series of grey regions, containing both the ebony and ivory master proteins. A gene might interpret this pattern such that it was only switched on in the regions of pure ebony or ivory, but not in the grey regions in between (that is, the gene would have both ebony and ivory binding sites in its regulatory region, but would be switched on when only one of these was occupied). This gene would therefore only be active in the 14 regions of pure colour (i.e. 7 of pure ebony and 7 of pure ivory). The activity of this gene may in turn lead to the production of a hidden colour, say hazel, leading to 14 hazel stripes. Another gene might interpret the zebra pattern differently, say coming on only in the grey regions, leading to stripes of tangerine.

Further elaborations, through scents and sensitivities, could further refine the pattern to give the precise hazel-mauve-tangerine pattern. This is essentially the way the expression pattern of the segment polarity genes is established, through interpretation and elaboration of

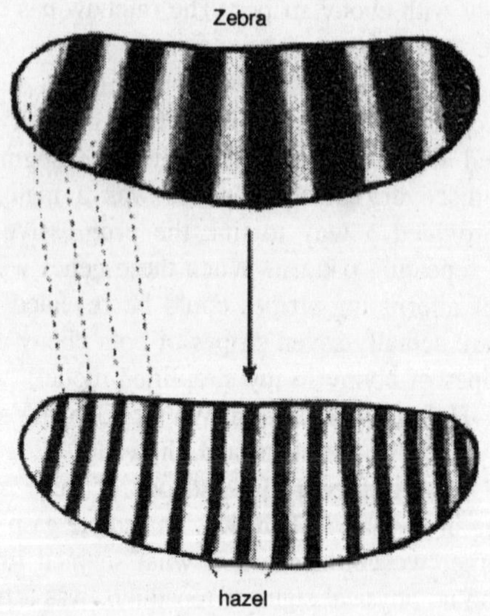

Fig. 4.8. Interpretation of zebra to give a pattern of 14 hazel stripes.

the zebra pattern established by the pair-rule genes. To summarize, two different patterns can be traced back to the rainbow. On the one hand, interpretation of the rainbow via the zebra pattern leads to the repeating pattern of hazel-mauve-tangerine stripes. On the other, the rainbow provides the frame of reference for the different types of green, providing variation between segments. That is to say, the rainbow pattern is interpreted along two routes. One gives a repeating pattern of hazel-mauve-tangerine stripes, the common theme that is depicted in every segment. The other route gives the progressive change in segment identity from the head end to the tail, providing a distinct green colour for each segment. (You may notice that the different greens cannot account for all the variation between segments, because there are fewer types of green than segments. Some of the variation between segments depends on other factors, such as the intensity as well as the type of green.) Although the two routes are shown as separate, they actually occur concurrently in the embryo as it develops,

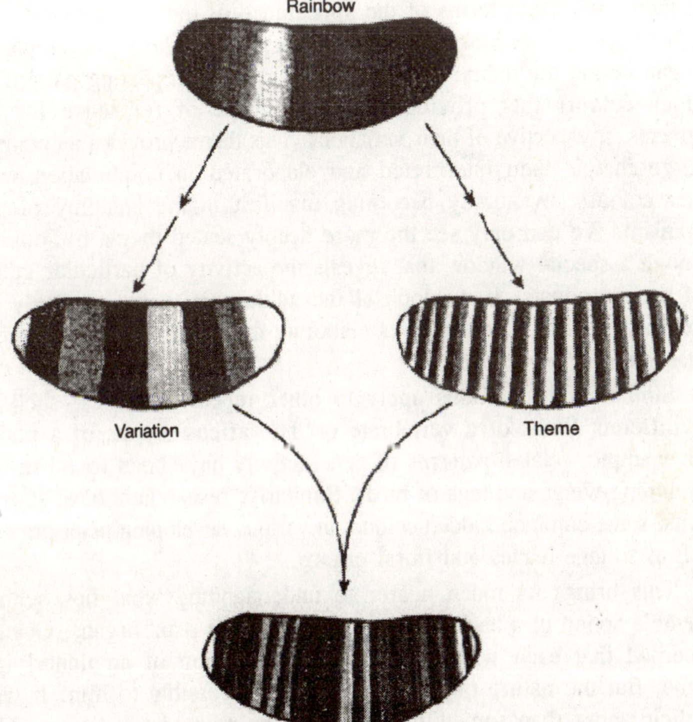

Fig. 4.9. Common source of theme and variations.

with some cross-talk between them that ensures theme and variation are kept in register with each other. The key point is that because both the variation and theme are an elaboration of a common fundamental pattern, the rainbow, their superimposition is automatically coordinated.

Implicit Themes and Language

We are now in a position to look again at implicit biological themes. By examining an adult fly, we get an overall impression of a repeating structure or theme, the segment, together with many variations upon it. Yet it is difficult to define the precise nature of this theme independently of the variations, because any individual segment will always be a combination of the two. If we try to define the theme explicitly, as a visible segment, we would have to say what sort of segment it was, but in doing so we would end up by describing only one particular variation, not the general theme.

There is no way out of this impasse so long as we try to define the theme purely in terms of the appearance of the organism. But as we have seen, if we look at how the segments arise during development we can define the theme in a different way: as a repeating pattern of hidden colours that provides an early frame of reference for all segments, irrespective of their variations. This theme provides a common pattern that is then interpreted and elaborated in combination with other colours, eventually becoming manifest in the anatomy of the organism. We can only see the more deeply seated theme by looking through a special window that reveals the activity of particular genes and their products. If we look at the adult, this theme can only be discerned implicitly, through its elaborate manifestation in the visible organism.

Similar considerations apply to other repeated themes, such as the different limbs of a vertebrate or the various organs of a plant. For example, related patterns of gene activity have been found in the developing wings and legs of birds. Similarly, researchers have started to find some common hidden colours in various developing plant organs, such as foliage leaves and floral organs.

This brings us much nearer to understanding what lies behind Goethe's notion of a leaf. By comparing different plant organs, Goethe discerned that each was a different manifestation of an underlying theme. But the nature of this theme was inaccessible to him. It was implicit rather than something that could be pointed to. As a result, his notion sounded mystical and obscure.

But we can now see that this follows automatically from the nature of development. The visible structures on a plant are based on the interpretation and elaboration of hidden patterns; patterns that can only be revealed by studying the activity of genes. In some cases, as with many of the foliage leaves on a plant, the same hidden patterns can be repeated and interpreted with little or no variation, so we end up with a series of very similar-looking organs. But in other cases, as with different types of foliage and floral organs, variation is superimposed on the repeating patterns so that they become manifest in different ways. This variation is itself another pattern of hidden colours, such as the various types of red that distinguish the whorls of flower organs. In these cases, repetitive themes may only be dimly perceived by looking at their visible manifestation in plant organs. That is to say, the theme is not a plant organ of any description; it is something that can only be dearly conceived of at a different level, in terms of processes that are normally hidden from view.

It seems to me that once this is appreciated, we can look at other types of implicit theme from a new perspective. Take the meaning of words, for example. What do we mean by a word like 'chair'? You might say that '*chair*' means an object with a seat, back and legs. But this would simply raise the question of what we mean by '*seat*', '*back*' and '*legs*'. Instead of one word, we now have to define the meaning of three. Each of these would in turn have to be described in more words, so we end up with an infinite regress. One way out of this problem might be to say that we know the meaning of '*chair*' not through a verbal definition but by forming a picture of it in our mind. If you were asked what a chair is, you quickly imagine what one looks like. But the problem here is that when you try to picture something in your mind, such as a chair, you always see one particular example, a certain type of chair. Yet the notion of a chair applies to all chairs, not just one, so how do we know the meaning of the word '*chair*' in general as opposed to individual cases? How can we imagine the shape that is common to all chairs, rather than one particular instance? The problem is the linguistic equivalent of Goethe's difficulty with defining the general notion of leaf, as the philosopher Ludwig Wittgenstein alluded to in his book Philosophical Investigations.

In other words, as soon as we try and define the meaning of the word '*leaf*' more precisely, by conjuring up a mental picture of one, what we see in our mind will be one particular leaf, not the general meaning. You might picture a simple broad fiat green apple leaf, or a leaf of grass, or a pine needle, but these are just examples, variations, not the general notion itself. Wittgenstein was using the example of a leaf to make a general point about language: whenever we try and conjure up a picture to represent a word that wc use, the picture will always be only one example rather than the general concept. Wittgenstein could have chosen any common object, such as a chair, to make the same point, but he probably chose a leaf because he saw the relationship between this problem and Goethe's ideas on plants.

Although Wittgenstein was concerned with the nature of the word '*leaf*' in the mental rather than biological sense, we can get a general insight into the problem he raised based on what we have learned in this chapter. We have seen that in the case of plants, the common unity of plant organs is not a visible leaf at all, but something that can only be described in other terms, such as an underlying pattern of hidden colours. By analogy, we might expect that common mental themes, our general notion of what '*leaf*' or '*chair*' means, could not be pictured in our mind's eye, but would lie hidden from us. In this case, the theme might reflect a common pattern of brain activity, rather than hidden colours. For example, every time we think of or use the term '*chair*', a common set of brain connections could be triggered, irrespective of the individual type of chair involved. This common pattern would itself have been established and elaborated upon during previous responses of our brain to the various situations in which the word '*chair*' was used, during which we eventually learned the meaning of the word. The important point is that we would be oblivious to this underlying pattern and would only be conscious of it indirectly, through its manifestation in particular cases, such as when we try to imagine what a chair looks like. The particular type of chair a person imagines may vary according to his or her state of mind, but all such chairs may nevertheless share some underlying patterns of activity in the person's brain. So although we have an implicit notion of a common meaning of the word '*chair*', we cannot access the underlying unity directly by introspection. As soon as we try to analyze it, by, say, visualizing something, we are stuck with a particular case rather than the underlying theme—much as when we look at a mature plant organ we see one particular manifestation rather than the common theme. It seems to me, then, that by getting a dear understanding of how repeated themes are elaborated during biological development, we can get a useful intuition into how analogous processes might operate in other cases, including the workings of our mind.

We have seen how both the idiosyncrasies and repetitions evident in the anatomy of organisms depend on hidden processes. Development involves the continual interpretation and elaboration of hidden colours, scents and sensitivities, giving an internal patchwork that underlies the final geometry we see. If one or more of these hidden components is lost through mutation, there is a corresponding change in the geometry of the organism that develops. But there is still the issue of how this process of internal painting eventually becomes manifest in the visible structure of each plant or animal.

5

HOME BOX GENES

As Gregor Samsa awoke one morning from uneasy dreams he found himself transformed in his bed into a *gigantic insect*. He was lying on his hard, as it were armour-plated, back and when he lifted his head a little he could see his dome-like brown belly divided into stiff arched segments on top of which the bed-quilt could hardly keep in position and was about to slide off completely. His numerous legs, which were pitifully thin compared to the rest of his bulk, waved helplessly before his eyes.

This is how Kafka starts his short story, *Metamorphosis*, describing the life of Gregor after he was transformed into an insect. Kafka's choice of an insect as the vehicle of his nightmare was not accidental. Insects live within their skeletons: they carry their supporting framework on the outside of their body, conjuring up a *claustrophobic image* of being helplessly trapped and entombed within a hard casing. They seem to have an imprisoned and alien existence, providing useful fodder for science fiction and horror stories. Nevertheless, there is a basic similarity between insects and ourselves. Like us, they have a head with mouth and eyes at one end, a body bearing limbs, and an anus at the tail end. This is what allows Kafka's transformation of a human into an insect to work so well: we can readily substitute human parts for corresponding parts of a beetle, head for head, main body for main body. We might take this as indicating that vertebrates, animals with internal bony skeletons such as ourselves, and arthropods, jointed animals with an outer casing such as insects, are formed on similar principles. Perhaps there is a common system that underlies the formation of these different types of animal. Another view would be that this similarity between vertebrates and arthropods is simply a trivial consequence of the limited number of ways that an animal can function. There are only so many ways that animals can operate. Having a head at one end, a main body with limbs, and a tail end, is a particularly convenient arrangement and it is not surprising to find it in different types of animal. In this view, a beetle and human are entirely different types of creature, any resemblance between them being a superficial consequence of limitations on the ways animals can function.

The question of whether vertebrates and arthropods are formed in a similar way or are built entirely differently was a burning issue in the early nineteenth century. It culminated in a famous confrontation in 1830 between two highly respected professors at the Muse d'Histoire Naturelleum in Paris: Georges Cuvier and Etienne Geoffroy Saint-Hilaire.

Form and Function

The early part of the nineteenth century was the golden era of comparative anatomy, with the Muse d'Histoire Naturelle in Paris being at the heart of many of the most exciting newum discoveries. Scientists were busily dissecting, describing and comparing many different types of animal for the first time, trying to discover the secrets of their internal anatomy. The legacy of this era can still be observed today in the Paris Museum's Gallery of Comparative Anatomy. There you can see what looks like a skeleton march: a stunning display of skeletons from all sorts of animals, arranged so that they all appear to be walking in the same direction. The skeletons bear witness to a period in which the study of anatomy was at the cutting edge of biological research. Based on these sorts of investigations, Cuvier and Geoffroy each thought they had uncovered the unifying principles that governed anatomy. Their approach was, however, very different.

You might classify musical instruments in two ways. One would be to classify them according to how they look: violins and cellos have a similar shape and are made of wood and string; trumpets and horns look similar and are made of metal. This is a classification based on the form or structure of the instrument. Alternatively, you could classify them according to how they make sounds: you have to pluck or bow a stringed instrument; a woodwind instrument is sounded by blowing, either directly or through a reed; to play a brass instrument you have to press your lips against a mouthpiece so that they vibrate when you blow; finally, you bang or hit a percussion instrument. In this case, the classification is based on the way the instrument functions.

These two approaches to classification—the form of the instrument or the way it functions—emphasize different aspects of the instruments. If you were to classify a *saxophone* based on form, you might place it with the brass instruments, whereas if function was your main Criterion you would place it in the woodwind section because you play it by blowing through a reed (saxophones are normally classified as woodwinds for this reason). In spite of these differences, the two types of classification will often give similar answers because the structure of the instrument is obviously closely connected to the way you make it sound.

For Cuvier, the key to understanding the structure of an animal lay in the way it functioned. Each animal was beautifully designed to function in a particular way and this dictated its form and structure. To Geoffroy, things were the other way round: the fundamental feature of animals was their unity of form. The specific way they functioned was a secondary matter. Following these two approaches, Cuvier and Geoffroy each arrived at a different formulation of the key rules underlying anatomy.

Cuvier and Functional Integration

Based on extensive animal dissections and studies, Cuvier decided that it was no good trying to understand organs or tissues in isolation: they only made sense when they were seen as parts of an integrated active individual. Why, for example, should birds have feathers? Feathers only make sense if they can be attached to a specialized type of forelimb, making a wing. A wing only makes sense if there is a certain type of collar- and breastbone for the wing muscles to attach to. The muscles in turn can only work if they can be provided with high levels of oxygen, requiring a particular type of chest and breathing system. Carrying on in this vein, the whole body plan of a bird could be deduced, starting from just a feather. Once told that an animal has a feather, we could work out that it must also have a certain type of collar-bone, and given a particular collar-bone we could infer that the owner had feathers. A feather without a bird, or a bird without feathers, just would not make any sense: the resulting animal would be illogical, unable to function properly and so could not exist. Cuvier applied and extended this principle to an enormous range of animals. Given a single fossilized bone he became expert at predicting what the rest of the skeleton looked like, what the animal ate and how it moved.

Cuvier was aware, however, that in practice many of the relations between parts had to be worked out retrospectively. If someone had never seen a bird and was given a feather, it is very unlikely that he or she would be able to deduce the structure of a bird from scratch. In practice, we cannot work everything out from first principles, but need to glean knowledge from the animals around us. Cuvier's deductions using fossilized bones were based on comparisons with bones of other known skeletons rather than purely logical deductions. He thought that this had more to do with our inadequacies in logical thinking than with any weakness in his theory. If we were intelligent enough, perhaps we could work out the structure of a bird from a feather without first needing to look at any examples.

It is a pity that Cuvier never met Sherlock Holmes, another great exponent of the power of deductive logic. This is how Holmes describes the art of deduction in A Study in Scarlet:

From a drop of water, a logician could infer the possibility of an Atlantic or a Niagara without having seen or heard of one or the other. So life is a great chain, the nature of which is known wherever we are shown a single link of it. Like all other arts, the Science of Deduction and Analysis is one which can only be acquired by long and patient study, nor is life long enough to allow any mortal to attain the highest possible perfection in it.

To Cuvier, the logical principle that connected the different parts of an organism together was their functional interdependence: parts were inextricably linked because they relied on each other to work properly. To some degree this may seem rather obvious and it had indeed been stated by others before. Cuvier's achievement was to pursue this idea in a systematic way and to elevate it to a fundamental law which he called the *Conditions of Existence*: it is this mutual dependence of the functions and the aid which they reciprocally lend one another that are founded the laws which determine the relations of their organs and which possess a necessity equal to that of *metaphysical* or *mathematical laws*, since it is evident that the seemly harmony between organs which interact is a necessary condition of existence of the creature to which they belong and that if one of these functions were modified in a manner incompatible with the modifications of the others the creature could no longer continue to exist.

For a creature to exist at all, its various parts must function together properly. This means that if one part was altered so that it no longer worked well with the others, the organism would not be able to survive and exist. Functional integration was therefore a necessary condition for existence and was the guiding principle behind all animal designs. Cuvier took the harmonious arrangements observed in animal anatomy to be a reflection of God's wisdom: species had been created by God along logical principles with parts that worked well together.

Perhaps Cuvier's crowning achievement was the way he applied his principle to the overall classification of animals. After surveying the animal kingdom, he proposed that there were basically four types of animal, each organized along different lines: (1) vertebrates, animals with a backbone, (2) molluscs, soft bodied animals such as slugs and snails, (3) articulates, jointed animals such as insects and shrimps (i.e. arthropods), (4) radiates, radially symmetrical animals such as starfish and jellyfish. Any animal could be classified as belonging to one of these four categories, or embranchements, just as an orchestral instrument can be classified as belonging to strings, woodwinds, brass or percussion.

Coming up with this four-fold classification may not seem particularly astounding, but imagine you had the task of dividing up the entire animal kingdom in a sensible way. Where would you start and what would your criteria be? You might begin with divisions like birds, fishes, mammals, etc. The trouble with this sort of classification is that it is somewhat superficial and *anthropocentric*: it is biased towards animals we are more familiar with. When set in the context of the whole animal kingdom, birds, fishes and mammals are seen to be all organized along very similar lines. Cuvier came up with a less parochial classification that recognized fundamental features of animal function and structure. To Cuvier, the four embranchements represented qualitatively different types of functional organization. In his system, mammals, birds and fishes all belonged to just one of the embranchements, the vertebrates. The remaining three embranchements were reserved for fundamentally different animals such as insects, slugs and starfishes. (This minority status of vertebrates is even more apparent in modern classifications which recognize about thirty-five major groups or phyla, of which the vertebrates are only one.)

Now according to Cuvier, although you could fruitfully compare animals belonging to the same embranchement with each other, comparisons between embranchements would be meaningless. It would be like trying to compare a violin with a trumpet. Humans, for example, belonging to the vertebrates, could not be compared with insects, belonging to the articulates, because they are organized in a totally different way. For Cuvier, the key to the animal kingdom was functional integrity, which he thought led inexorably to the fourfold classification of distinct types. It was this division of animals into qualitatively different categories that eventually led to Cuvier's conflict with Geoffroy.

Geoffroy and the Principle of Connections

Geoffroy approached the problem of animal structure from a quite different angle. Compare the way horses and humans walk. When a horse walks, the back legs appear to bend at the knee in such a way that the lower/ part of the leg swings forwards and up. Try to do a similar thing with your legs and you'll end up in hospital because when you walk, the lower part of your leg swings backwards below the knee. The front legs of a horse seem to be a bit more comfortable to think about: the lower part of the leg swings back below the knee. However, we should really be comparing the front legs of a horse to their real human equivalent, arms. If you let your arms fall to your side and bend them at the elbows, the lower arm swings forward, the opposite to the way a horse's front legs seem to bend.

To see why our joints appear to work in the completely opposite way to a horse's, we need to look at the internal anatomy of limbs. If you follow the arrangement of bones in your leg starting from the

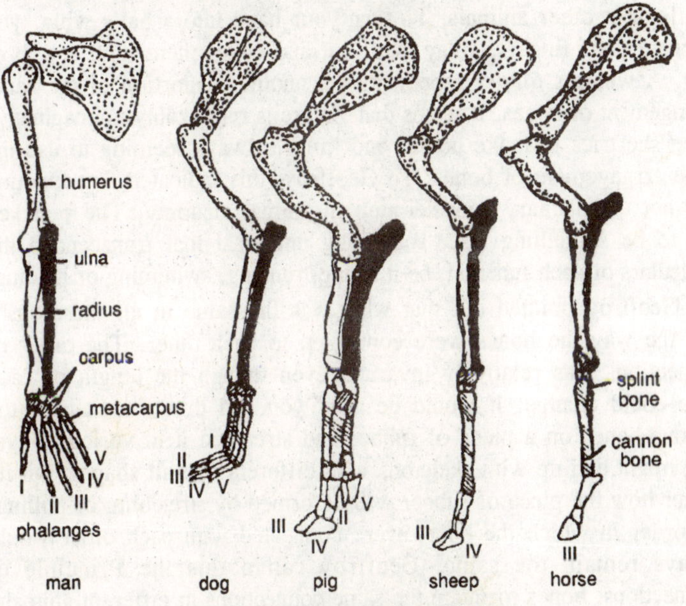

Fig. 5.1. Mammalian forelimbs

pelvis, you have a single thigh bone (femur), which is followed at the knee joint by two shin bones (tibia and fibula), which are in turn jointed at the ankle with bones in the foot. The back leg of a horse has a very similar arrangement but if you try to match its bones with ours, one for one, it turns out that what I have been calling its knee corresponds to our ankle, and its lower leg matches the middle bones (metatarsals) of our foot, ending in the middle toe. To turn yourself into a horse, you have to imagine the middle bones of your foot growing to the length of your shin, and the nail on your middle toe enlarging to form a hoof. The same is true for the front limbs: you have to imagine the bones of your middle hand and finger growing to be about as long as your forearm. It is now easy to move like a horse. You swing the back legs forwards by bending your ankles and the front legs backwards by flexing your wrists. The sequence of bones in a horse and the way they bend at the joints is similar to ours, it is just that the relative lengths of the various bones are quite different.

Now surely if you were designing horses and humans purely on functional grounds, it is unlikely that you would choose exactly the same arrangement of bones and be satisfied with just varying their relative lengths. Why not have different arrangements in each case that best suit the purpose they are put to? The problem gets worse as you look at other animals. To turn your hand into a bat's wing, you elongate your fingers, splay them out and cover them with a web of skin. A whale's *flipper* represents yet another distortion of the same arrangement of bones. It seems that Nature is remarkably unimaginative when she tries to make bodies and limbs, always seeming to use the same arrangement of bones. To Geoffroy, this indicated that function was not the primary consideration in animal anatomy. The real key had to be something else: something universal that transcended the particulars of each function, be it flying, running, swimming or holding.

Geoffroy pointed out that what was the same in all these cases was the way the bones were connected to each other. The order of connections was relatively invariant even though the length of each bone could change. It would be as if you had drawn a map of the various bones on a piece of rubber and stretched it in various ways. You might end up with skeletons with different overall shapes, but no matter how the piece of rubber was deformed by stretching or pulling, the order in which the bones were connected with each other would always remain the same. Geoffroy called this the Principle of Connections: bones maintain the same connections in different animals, irrespective of which particular functional use they are put to. In the 'Who's Who' of bones it is not what you do but who you are connected to that matters.

It followed from this that all mammals had a common map of bone connections: they all conformed to a unity of plan. This unified plan did not correspond to any particular animal but to a basic layout that was common to all. In a similar way, it has been tried to illustrate all stringed instruments by a generalized diagram showing their basic layout. The generalized diagram is not itself an instrument, it is a map that depicts the common arrangement of parts in a violin, cello or double-bass. The unified plan is an abstraction, a general property which is manifested in individual cases.

Conflict and Controversy

Within certain limits, Geoffroy's principle of connections and unity of plan were acceptable and even welcomed by Cuvier. It clarified the well-known fact that different mammals display similar bone arrangements. It was only when Geoffroy started to claim true universality and primacy for his principles that trouble started to brew. First of all, he claimed that mammalian skeletons were not only similar to each other but also displayed the same basic connections as skeletons of fishes. On the face of it, this seemed absurd because there were many fish bones, particularly those in the head, that had no obvious counterparts in mammals. Undeterred by this, Geoffroy applied his principle of connections, ignoring superficial differences in the relative sizes and functions of bones, to establish what he thought were the more fundamental similarities.

Cuvier found some of Geoffroy's comparisons convincing, others less so, and if matters had stopped there, their disagreements might merely have been a question of emphasis. Geoffroy's next step was to change all that. In 1820 he proposed that not only did different vertebrates, such as mammals and fishes, have the same type of bone arrangements, but that they also shared the same basic layout with insects— animals that don't even have any bones! This flatly contradicted Cuvier's view that vertebrates and insects belonged to distinct embranchements. To Cuvier, comparing a human to a fly was simply not allowed: it was like comparing chalk with cheese.

Geoffroy supported his radical claim by pointing out that insects and vertebrates shared many features in their basic organization. They both had a head with mouth and eyes, and the rest of the body was divided into two segmented regions: the thorax (the chest in humans) and abdomen. The main difference between them was that whereas vertebrates were supported by a bony skeleton from within, insects were supported from the outside by a hardened casing. Geoffroy proposed that these apparently different types of organization were really two sides of the same coin; it was just that vertebrates were organized around a skeleton whereas insects lived within theirs. To turn yourself into an insect you have to imagine your skeleton expanding outwards, whilst your body remains the same size. Eventually the bones reach and enclose the skin, and all the soft tissues and vital organs end up interior to them, like a very elaborate bone marrow. He believed that once this was appreciated, detailed correspondences between insect and vertebrate structures fell into place according to his principle of connections. For example, the main trunk of the insect would correspond to the vertebral column (backbone), and its legs would be equivalent to our ribs. Not surprisingly, this view of insects as perambulating skeletons walking around on their ribs was greeted with disbelief by many: the detailed correspondences that Geoffroy drew between insects and vertebrates just seemed too fanciful. Nevertheless, Geoffroy stuck to his guns, convinced that the principle of connections revealed the deep unity shared by organisms. Sceptics, he thought, were simply being distracted by the more obvious superficial features.

For Cuvier, these comparisons between vertebrates and insects were both ridiculous and offensive, and for the next ten years he tried to undermine Geoffroy and his followers. As the historian Toby Appel summarizes: the tension between the two naturalists was apparent to all, and weighed heavily not just upon the principals but upon the entire scientific community. In all their lectures, articles, books and reports at the Academie, Cuvier and Geoffroy continued to snipe at each other, but neither responded to the other head-on. Geoffroy accused Cuvier of denouncing him behind his back, rather than presenting a reasoned critique of his work. For ten years, beginning in 1820, he tried to lure his scientific opponent into a public confrontation, but Cuvier preferred the strategy of indirect attack.

Then, on February 15th, 1830, the *bubble burst*. The occasion was one of the weekly meetings of the Academie des Sciences in Paris where Academicians presented reports or papers. Two relatively inexperienced young naturalists, Meyranx and Laurencet, had submitted a memoir to the Acade pointing out that if a vertebrate was bent back on itself, its organs would be in similar positions to those of a cuttlefish, a mollusc. This was music to Geoffroy's ears because it implied that his universal type could now be extended to yet another of Cuvier's embranchements: the molluscs. When Geoffroy was asked to give a verbal report to the Acade on this memoir, hemie seized the opportunity not only to commend it but also to promote his own ideas. Surely the memoir illustrated how, by applying the principle of connections, real progress could be made to reveal the fundamental unity of animals. This was so much more exciting than the old-fashioned view, which he exemplified by a quote from one of Cuvier's classic works.

This was too much for Cuvier. He stood up at the end of Geoffroy's report and protested against both the scientific content and the personal slur in the report, promising to deal with the issues in detail at a later date. At the following week's meeting of the Academie, Cuvier arrived armed with diagrams of a cuttlefish and a vertebrate bent back on itself, prepared to do battle. He proceeded to show that many of the proposed similarities between molluscs and vertebrates simply did not hold up to close scrutiny. The debate, which was now starting to attract considerable public attention, continued vigorously for four more meetings, with the adversaries reduced to squabbling at one point over who should be allowed to speak first. The last of these meetings signalled the end of the verbal contest before the Academie, but by then it had caught the imagination of the scientific community and the public at large. Here was a fundamental issue about our place in Nature being disputed between two giants of the French intellectual establishment. Newspapers, journals and books publicized the debate, some siding with Cuvier, others supporting Geoffroy.

Darwinian Solution

About thirty years after the debate in the Academie, Charles Darwin thought he had provided a solution to the conflict in his book On the *Origin of Species* (1859). To Darwin, anatomical arrangements were simply a consequence of the way evolution worked. The functional aspects of anatomy that Cuvier emphasized could be accounted for as adaptations resulting from natural selection: organisms were made of parts that worked well together because such an arrangement was more likely to lead to survival and reproduction. In Darwin's words:

> The expression of conditions of existence, so often insisted on by the illustrious Cuvier, is fully embraced by the principle of *natural selection*. For natural selection acts by either now adapting the varying parts of each being to its organic and inorganic conditions of life; or by having adapted them during long-past periods of time.

To Darwin, functional integration was an adaptation that had gradually evolved over millions of years by selection. He acknowledged

Cuvier's principle but instead of seeing it as a reflection of God's design, he explained it as an outcome of natural selection. Unity of form, as highlighted by Geoffroy, could also be accounted for by Darwin through common descent. The anatomy of horses, humans and all other mammals was similar as a consequence of their shared ancestry. Because evolution proceeded by gradual alterations, each step modifying what went before, ancestors and their descendants would be expected to retain certain features in common. In the case of vertebrates, Darwin pointed out that the sizes and shapes of bones were more likely to change than their arrangement, accounting for Geoffroy's principle of connections:

The bones of a limb might be shortened and widened to any extent, and become gradually enveloped in thick membrane, so as to serve as a fin; or a webbed foot might have all its bones, or certain bones, lengthened to any extent, and the membrane connecting them increased to any extent, so as to serve as a wing: yet in all this great amount of modification there will be no tendency to alter the framework of bones or the relative connexion of the several parts.

Fins, wings and other types of mammalian limb had a similar set of bone connections because they evolved by gradual modification from a common ancestral appendage. Geoffroy's principle of connections and unity of plan were no longer fundamental laws but simply a historical consequence of common ancestry.

There was a further sting in the tail for Geoffroy's principle. Because the basic layout of an organism was featured in a common ancestor, this feature itself would be something that had evolved. For example, the common ancestor of mammals did not appear from nowhere: it also had an ancestry going even further back in time. Thus, according to Darwin, the layout that linked a class of animals, such as all mammals, was itself an *adaptation* that had arisen earlier on in their ancestry by *natural selection*. This meant that the unified plan (or unity of type, as Darwin referred to it) had its origins in the *functional adaptations* of the past. For Darwin, Geoffroy's law of form became subsumed in Cuvier's law of function: It is generally acknowledged that all organic beings have been formed on two great laws—*Unity of Type*, and the *Conditions of Existence* . . . in fact, the law of the Conditions of Existence is the higher law; as it includes, through the inheritance of former adaptations, that of Unity of Type.

Now if a basic layout is something that has arisen during evolution, you could imagine its having arisen more than once. For instance, the common ancestor of vertebrates could have evolved its basic layout quite independently of the common ancestor of arthropods (insects, crustaceans, etc.). There was no fundamental reason for supposing that layouts of insects and vertebrates had anything in common. Of course, if you go back far enough in evolutionary time, you might come across the ancestral stock from which both vertebrates and insects were descended. But this ancient organism might not have had much of a layout to speak of: it may have been a simple aquatic animal that had neither an internal nor an external skeleton, and therefore would lack a framework of parts that could be related to those of modern insects or vertebrates. And even if there was some sort of simple layout in this ancient animal, any vestiges of similarity between this and modern forms would most likely have been wiped out during the long periods of evolution. Given the many notable differences between the anatomy of insects a vertebrates, it seemed most likely that much of their basic layout had effectively arisen independently and had little in common. Modern insects and vertebrates were constructed along fundamentally different lines and could not be meaningfully compared. As far as this issue was concerned, Cuvier had won the day and Geoffroy's position became no more than a historical curiosity.

Vertebrate Homeobox Genes

The first stirrings that were to rekindle the issue of how vertebrates were related to arthropods came in 1984, about one hundred and fifty years after the original dispute between Cuvier and Geoffroy. It was the time when the identity genes affecting segments in the fruit fly had just been isolated. Recall that these genes are needed for a set of hidden colours that define distinct regions in the fly. The genes code for a set of related proteins, symbolized by eight green colours, distributed from head to tail in alphabetical order: from apple-green at the head end to herb-green at the rear.

We might say that the arrangement of these hidden colours sets a basic layout for the fly from head to tail. Remember that the main body of a fly consists of a series of 14 segments with different identities: three head segments bearing appendages such as antennae, followed by three segments of the thorax bearing legs and sometimes bearing wings as well, and then eight distinct segments in the abdomen. These distinctions between the segments, the head to tail layout, depend on a family of green hidden colours. *Mutations* that remove one or more of the hidden colours result in a failure to distinguish between some regions, giving several segments with the same identity. If all the hidden green colours were missing, you would end up with a monotonous arrangement of 14 similar segments, losing the distinctive layout from head to tail. In other words, the anatomical layout itself depends on a deeper layout: the map of hidden colours.

Given the many differences in the anatomical layout of insects and vertebrates, we might expect that their map of hidden colours would also be quite distinct. This issue could be examined directly as soon as the genes for hidden colour, the identity genes, had been isolated from flies. Once you ha isolated a gene from one species, it is possible to look for related genes in other species. So, shortly after the identity genes needed for hidden colours had been isolated from flies, Bill McGinnis and Michael Levine, in Walter Gehring's lab in Switzerland, started to investigate whether similar genes were present in other animals. Because of their distinctive anatomical layout, the expectation was that vertebrates would either not contain similar genes at all, or that they would have a different significance for vertebrates than for arthropods.

The method they used to look for *related genes* in other species is not too different, in principle. An *isolated gene* is used to make a molecular probe, but in this case the probe does not detect RNA or protein molecules, but a specific stretch of DNA sequence (the probe is itself a DNA molecule that has been chemically modified). Each organism contains many thousands of genes, each gene being a stretch of DNA. The probe can recognize a particular DNA stretch in this vast array: it sticks or hybridizes only to this stretch, distinguishing it from the others. The probe that McGinnis and Levine made was derived from an identity gene of fruit flies and was used to recognize any stretch of DNA that was similar to it in sequence. If a similar or related gene happened to be present in the DNA of a different species, such as that of a human, the probe would locate or cross-hybridize with that gene. They tested their probe on DNA from a range of other organisms, including worms, as Bill McGinnis recalls: Mike suggested we might as well test some worms, so my wife Nadine walked to the local fisherman's bait shop and got a variety of insects and segmented worms. The DNA isolated even before we identified what species they were, and put in some vertebrate DNAs hoping for some cross-hybridization, but really thinking of them as negative controls. At the time of first blot out, there were obviously some strongly hybridizing fragments in the human, frog and calf lanes.

The probe was recognizing or *cross-hybridizing* with stretches of DNA from vertebrates, indicating that these animals had genes that were similar or related to the identity gene from flies. McGinnis repeated the experiment and consistently got the same answer, so that eventually people became more convinced. Now as soon as genes from other species have been identified by a probe in this way, they can be isolated or fished out and studied in detail. Several laboratories therefore embarked on a series of molecular fishing expeditions, using probes based on the fly *identity genes* to isolate similar genes from vertebrates including frogs, mice and humans.

The vertebrate genes isolated in this way were remarkably similar in many details to the genes from flies. The first point of similarity came from comparisons of DNA sequence. Recall that the eight genes affecting segment identity in flies arose by gene duplication and divergence. Consequently, they all share a very similar stretch of sequence in their DNA called the *homeobox*, and because of this, these identity genes are sometimes referred to as *homeobox genes*. The homeobox is in the coding region, so these genes produce related proteins, symbolized by a family of green colours. Now the various genes isolated from vertebrates also contained a homeobox:s they were also homeobox genes. This meant that the vertebrate genes coded for proteins that belonged to the same family as in flies: the family of green colours. This is perhaps not so surprising, as the vertebrate genes were after all isolated because of their similarity to the fly genes. The probes were designed to find genes in vertebrates that were related to the identity genes of flies, so it is not too unexpected that genes isolated from vertebrates in this way would contain similar stretches of DNA, such as the homeobox. What was more intriguing was the extent of correspondence between the different types of fly and vertebrate homeobox genes.

Not only did the vertebrates contain genes for a family of green colours, but each type of green was also represented. For example, one of the vertebrate homeobox genes was most similar to the fly homeobox gene needed for the *apple-green* colour: its DNA sequence was clearly nearer to the gene for *apple-green* than to any of the other seven homeobox genes in the fly. Another vertebrate homeobox gene was most similar to the fly gene producing *bottle-green*. Yet another was obviously closest to the gene for *cyprus-green*. Each vertebrate gene could be matched up as a counterpart to one of the eight fly homeobox genes. Thus, vertebrates had a series of homeobox genes that produced the equivalent of each type of green, from *apple-green* to *herb-green*.

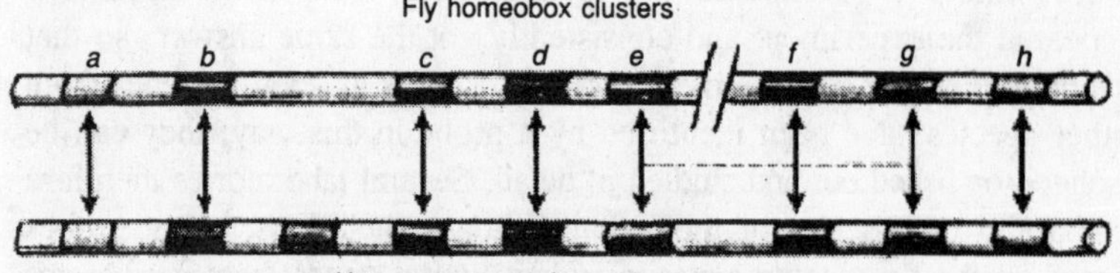

Fig. 5.2. Similar arrangement of homeobox genes in fruit flies and humans or mice.

The next point of similarity between fly and vertebrate genes came from looking at how the genes were arranged in the DNA. In flies, the eight homeobox genes—coding for the green hidden colours—are clustered together along the DNA. Five of the genes are in one duster and three are in a separate cluster. Each gene being labelled with the initial letter of its corresponding hidden colour. Thus, genes labelled a-e are in one cluster and f-h are in the other. As the homeobox genes from vertebrates started to be analyzed, it became clear that they were also clustered in a very similar way. Robb Krumlauf, who was working in London on the homeobox genes from mice, recalls:

Walter Gehring in late '86 had the first homeobox workshop. There were about eight of us guys working on vertebrates sitting there talking to each other. At that time we all had small groups of individual homeobox genes. Where the difference came, it is that Denis Duboule, who was in Pierre Chambon's lab in Strasbourg at that time, had a cluster of four or five genes like my duster and we started talking to each other and trading sequence.

It was becoming apparent that the various homeobox genes in vertebrates were clustered together in a particular way. The arrangement of the homeobox genes in mice was then pieced together in detail by Robb Krumlauf and colleagues in London, and Denis Duboule and Pascal Dolle in Strasbourg. At about the same time, Eduardo Boncinelli and colleagues in Naples were working out the arrangement of human homeobox genes. The results from mouse and humans were essentially the same. There was a striking similarity in the way vertebrate and fly homeobox genes were organized: the genes were clustered together in a similar order in the DNA, from a-h. In one way, the vertebrate arrangement was even simpler than that in flies because the complete set from a was clustered in a single region of DNA rather than in two separate-h ones.

The strong implication of this detailed correspondence was that insects and vertebrates had inherited their clusters of *homeobox genes* from a *common ancestor*. The ancestral aquatic animal that eventually gave rise to both the vertebrates and insects must have had a set of homeobox genes in its DNA arranged in the order a These genes had then been preserved in pretty much-h. The same arrangement throughout the evolution of vertebrates and insects. At some time in the evolutionary lineage that gave rise to flies, the cluster became split into two separate parts to give the two clusters a-e and f-h (this split has not happened in all insect lineages because some insects, such as beetles, still have a single cluster).

There are some additional complications. First of all, mammals have four copies of each gene cluster in their DNA. It seems as though a long stretch of DNA, including the entire cluster of homeobox genes from a-h, has been duplicated several times during the evolution of vertebrates, resulting in four copies of each cluster (giving a total of about 32 genes). Following these large-scale duplications, there has been some divergence between the four clusters, giving rise to slightly different versions of each type of gene: a human or mouse contains several slightly different versions of the gene for apple-green, several versions for bottle-green, etc. Secondly, there have been some *extra duplications* (and some *deletions*) of individual genes within clusters. This means that some clusters can have more genes than others and that some of the vertebrate genes cannot be matched up unambiguously with their counterparts in flies. Even with these qualifications in mind, the degree of conservation between homeobox genes in insects and vertebrates is striking.

Hidden Map of Vertebrates

What significance might these homeobox genes have for vertebrates, given that the anatomical layout of vertebrate animals seems so different from that of insects? In flies, the homeobox genes are required to establish distinct territories in the animal, underlying the basic layout from head to tail. It could be that these genes play a comparable role in vertebrates, even though their anatomy looks so different from that of insects. Alternatively, the vertebrate homeobox genes might have nothing to do with the head-to-tail layout in vertebrates: they could have an entirely different significance for humans than for flies.

A good way to test these possibilities would be to look at when and where the homeobox genes are expressed in developing animals (recall that by expression of a gene I am referring to where it is switched on, producing its protein product, at any given time). In flies, each homeobox gene is expressed in a particular region along the head-tail axis of the early embryo. This gives a set of green hidden colours distributed in a particular order from head to tail, with apple-green at the head end through to herb-green at the tail end. If the vertebrate homeobox genes were also involved in the head-to-tail layout, they might also be expressed in a similar order of territories in the organism. To examine this, vertebrate embryos were stained to reveal where the various homeobox genes were being expressed: where they were on or off.

Figure illustrates human embryos at various stages of development, from the second to the fifteenth week after fertilization. They are shown at about their natural size. You can see that there is a distinct head and tail end to embryos even at the earliest of these stages. As development proceeds, various structures, such as the limbs, gradually appear. All mammalian species follow a very similar course of early development, although some may go through the stages more rapidly than others.

Now look at two mouse embryos stained to reveal the expression of particular mouse homeobox genes at an early stage of development. One of the genes codes for an equivalent of the bottle-green protein and the other for *cyprus-green protein*. You can see that each of these genes is expressed in a discrete region of the embryo, starting at a

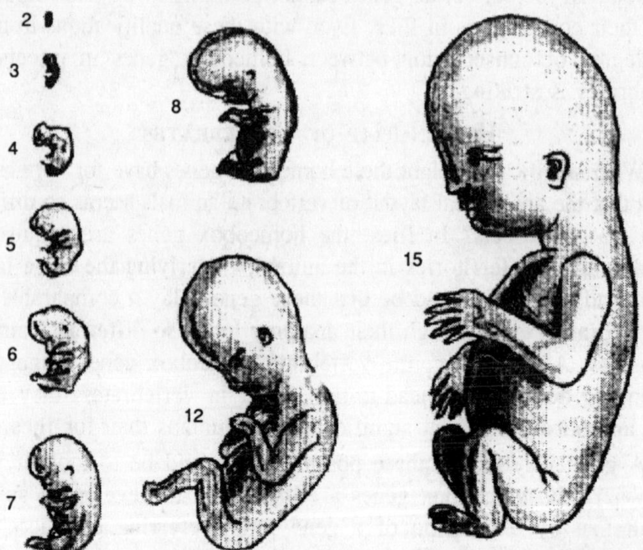

Fig. 5.3. Human embryos from the second to the fifteenth week, natural size. Age in weeks is indicated next to each embryo.

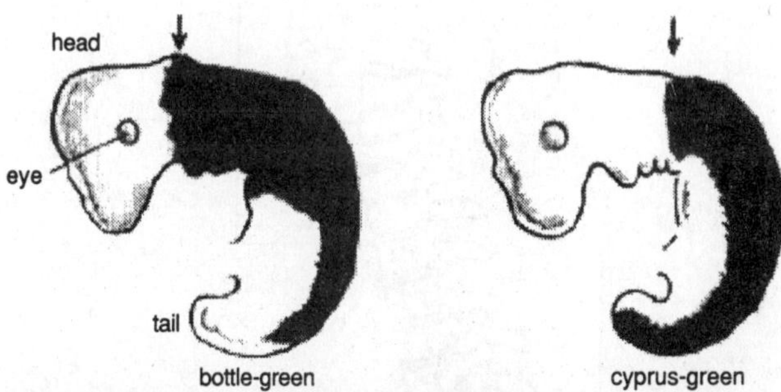

Fig. 5.4. Diagrams representing mouse embryos stained to reveal the expression pattern of a gene needed for bottle-green (left) or cyprus-green (right). Note that the region of cyprus-green starts further back than bottle-green.

fixed distance from the head and extended back from this. Most importantly, the starting point for the bottle-green colour is nearer to the head than *cyprus-green*. That is, the head-to-tail order is the same as in the fly: bottle-green is ahead of *cyprus-green*. Similar experiments with all the other homeobox genes gave comparable results: the order of expression of the genes in vertebrates was the same as in flies, with apple-green being nearest the head end, followed by bottle-green, then cyprus-green, and eventually ending up with herb-green towards the rear. Together, the various greens gave a series of distinct regions going from head to tail, much as had been observed for fruit flies.

This was a startling result because it showed that there was a common aspect to the organization of insects and vertebrates: they share an underlying map of colours that follow each other in the same relative order from head to tail. It is reminiscent of Geoffroy's principle of connections, but here it applies to hidden colours rather than bones. Whereas Geoffroy's rule was based on the order of parts of the skeleton, in this case it is the order of hidden colours that is held in common. Vertebrates and insects are united by a set of green hidden colours that are connected or ordered in a similar way from head to tail.

There is still the question, however, of what this map of green hidden colours signifies for vertebrates. In flies, they establish territories in the animal from head to tail, conferring distinctions between the various segments. Take away one or more of the colours, through mutations in the genes, and the identities of some segments are no longer distinguished. Perhaps the hidden colours also provide distinctions between the different parts of vertebrates. Although vertebrates are not segmented in the same way as insects, they are made up of repeated elements of various types from head to tail. This is most obvious in

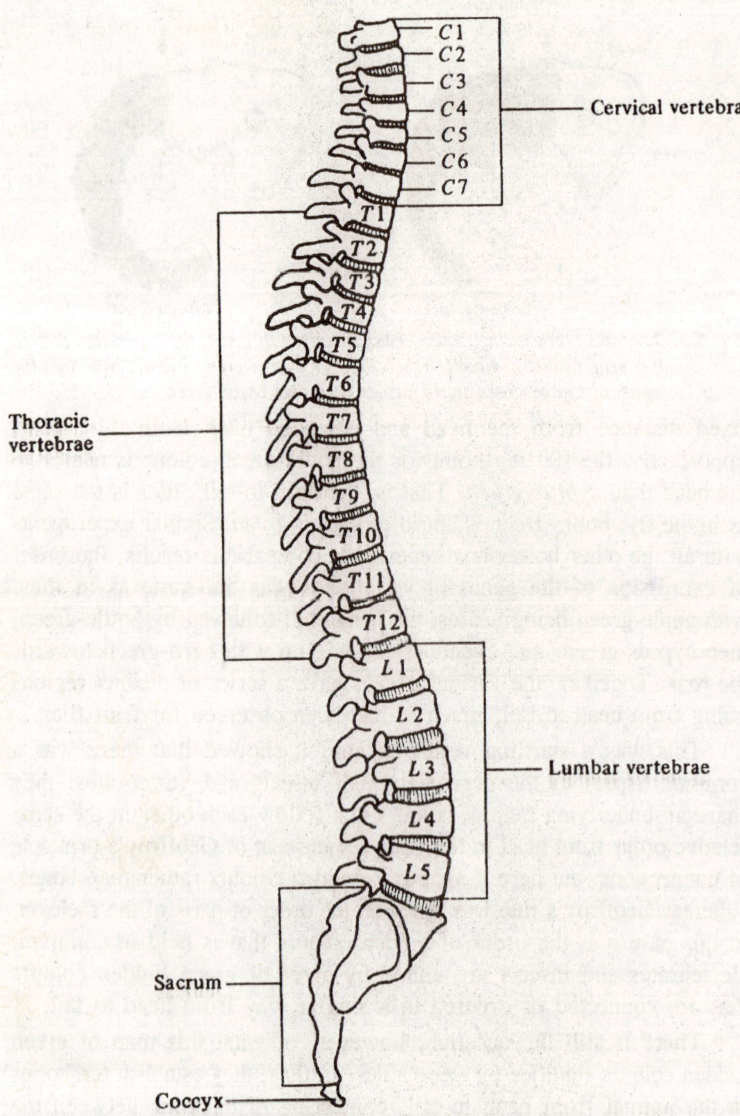

Fig. 5.5. Vertebral column: Lateral view. Individual vertebrae are numbered; vertebral regions are indicated.

the backbone, which is made of a series of repeating units or vertebrae. Going from head to tail in humans there are seven vertebrae in the neck region (*cervical vertebrae*), followed by twelve vertebrae in the chest region, bearing ribs of various lengths (*thoracic vertebrae*), followed by another five vertebrae without ribs in the abdominal region (*lumbar vertebrae*). Beyond this, there are several other vertebrae at the base of the back. In mammals with tails, the vertebrae can continue much further than this. The various types of vertebrae from head to tail are the most obvious place to look for a comparable role of the green hidden colours in vertebrates: perhaps the green colours provide distinctions between vertebrae from head to tail, similar to the way they distinguish between insect segments.

If the hidden green colours are important in providing distinctions between vertebrae, mutants that lack one of the colours might be expected to show alterations in the identity of particular vertebrae. This could be tested in mice because a method had been developed that allowed specific mouse genes to be inactivated by mutation. This method was used to inactivate particular homeobox genes, producing mutant mice that lacked the corresponding hidden colours. Some of these mutants did indeed show striking changes in the identities of specific vertebrae. An example taken from the work of Herve Le Mouellic, Yvan Lallemand and Phillipe Brulet, working in Paris in 1992. As in humans, mice have a series of thoracic vertebrae, each bearing a pair of ribs, followed by a series of lumbar vertebrae that do not have ribs attached to them.

The diagram on the right a mouse which has a mutation in one of its homeobox genes, a gene contributing to the *grass-green* hidden colour. In this mutant animal, the *first lumbar vertebra*—normally lacking ribs—has a pair of small ribs attached to it. The lumbar vertebra is displaying a feature, a pair of ribs, normally associated with the *thoracic vertebrae* that lie ahead of it in the backbone. Indeed, given that thoracic vertebrae are defined as those that bear ribs, we might say that the first lumbar vertebra has been replaced by a thoracic vertebra: its identity has changed from lumbar to thoracic. In other words, as with flies, the hidden colours are needed to provide distinctions along the *head-tail axis*, in this case between vertebrae. Mutations giving a loss of colour lead to a lack of distinction, such as a lumbar vertebra assuming a similar identity to a thoracic vertebra. You will notice that the transformation is not complete: the extra set of ribs is much smaller than the ribs that come off a normal thoracic vertebra. This is most probably because there are four copies of the *homeobox gene* cluster in mice, which can act to some extent as backups for each other. Mutation in a gene from one duster will, therefore, not completely eliminate a hidden colour, such as grass-

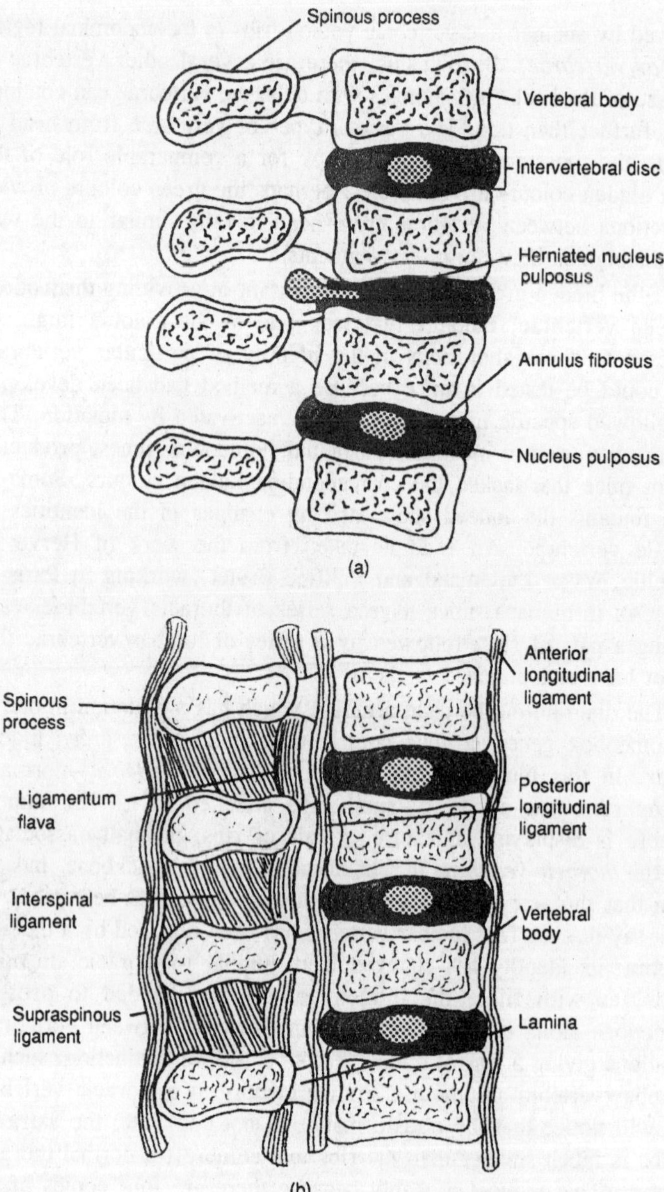

Fig. 5.6. Sagittal section of vertebral column.

green, because there are genes in the other dusters that can act to some extent as substitutes.

The analysis of mutations in other homeobox genes gave similar results. Mutant mice that lacked a homeobox gene contributing cyprus-green had alterations in their cervical vertebrae, resulting in the second neck vertebra assuming an identity more like that of the first vertebra.

More emphasis was given on vertebrae because they are the easiest feature to look at with respect to the head-to-tail axis of vertebrates, but the hidden green colours also affect other distinctions along this axis. In other words, the set of green hidden colours provide distinctions between territories from head to tail that will give rise to various structures during development, of which the vertebrae are the simplest example. To summarize, as with flies, the green hidden colours of vertebrates are needed to provide distinctions between various regions from head to tail of the animal. Loss or reduction in one of these hidden colours results in a lack of distinction between regions along the head-tail axis, most easily seen as some vertebrae that are normally quite distinct starting to assume similar identities. Thus, the anatomical layout from head to tail in vertebrates depends on an underlying map of green hidden colours similar to that found in insects.

From Colour to Anatomy

If insects and vertebrates contain a similar set of green hidden colours, how come they end up looking so different from head to tail? Recall that the hidden colours do not correspond to a set of instructions that specify how a structure should be made, like the construction of a particular type of segment: they simply provide a frame of reference that is then interpreted by genes. So although two organisms may have a similar map of green hidden colours, the way this is interpreted and eventually becomes manifest in the anatomy of the animals can be very different.

We are still very far from understanding the precise relationship between hidden colours and the final anatomy of any animal, be it a vertebrate or an insect. Nevertheless, it is tried and outlined some of the contributing factors to give a better idea of why these animals can end up looking so distinct, even though they share a similar underlying map.

We saw that hidden colours correspond to master proteins that can bind to particular sites in the regulatory regions of interpreting genes. The emerald-green master protein, for example, may recognize one short sequence of DNA, an E-site, whereas the grass-green protein recognizes a slightly different sequence, a G-site. By binding to these sites, the master proteins can influence whether an interpreting gene is on or off. In the simplest scenario, an interpreting gene with an E-site in its regulatory region will be switched on wherever the emerald-green protein is to be found in the organism. Similarly, a gene with a G-site will be expressed wherever the grass-green protein is located. Now although vertebrates and insects have a common map of green master proteins along the head-tail axis, many of the interpreting genes that respond to each master protein might be different in each case. We can imagine, for example, that many of the interpreting genes with an E-site in insects may be quite different from those with an E-site in vertebrates. During the long evolutionary time that has separated vertebrates from arthropods, the sites in the regulatory regions of interpreting genes could have changed. This would mean that the interpreting genes expressed in the emerald-green protein territory in vertebrates would differ from those expressed in the emerald-green territory of insects. The map of colours may be the same but their interpretation would differ.

The situation is, of course, much more complicated than my simplified scenario implies. For one thing, interpreting genes have several binding sites in their regulatory regions, allowing them to respond to a combination of hidden colours in the organism. There are many other types of master protein, in addition to the green family, each contributing a hidden colour to give a complex overlapping patchwork. The response of a gene to one or more green colours has to be set in the context of these other hidden colour patterns, some of which might be similar between vertebrates and insects, whereas many others might differ. Furthermore, the proteins encoded by the interpreting genes may themselves have evolved so that their effects on the organism can be different in insects and vertebrates. The way the interpretation of the hidden colours eventually leads to the visible structures of the animal is a very complex affair that can differ between insects and vertebrates in numerous ways.

All of this means that the process by which a set of hidden colours is interpreted and eventually becomes manifest in the final anatomy of the organism is likely to change in many ways during evolution. In insects, the green hidden colours eventually become manifest in the various types of segments, whereas in vertebrates their effects become apparent in other structures, such as the types of vertebrae. So even though the head-to-tail layout of insects and vertebrates depends on a similar map of hidden colours, you might never guess this on the basis of their anatomy alone.

A Hidden Unity

We can now take another look at the debate between Geoffroy and Cuvier. Cuvier's emphasis on function led him to see insects and vertebrates as being constructed on entirely different principles. There was no meaningful way of comparing these two types of animal. Geoffroy on the other hand was struck by the similar arrangements of parts in groups of animals, such as the mammals, irrespective of what particular function the parts served. Having revealed what he believed to be an underlying law of similarity, the way the bones were connected together, he tried to extend this law to other animals, going as far as insects. To do this he had to establish some sort of correspondence between the parts of an insect and a vertebrate. His solution was to match the outer skeleton of insects with the inner skeleton of vertebrates. Once this was done, detailed correspondences could be drawn, such as between the vertebrate backbone bearing ribs and the outer body of the insect bearing legs. Both systems could be seen to fall under the same umbrella of connections.

With the benefit of hindsight, we can see that there were some merits to Geoffroy's view. There is an underlying set of connections that is similar between insects and vertebrates: the map of hidden colours. Both types of animal have a family of green hidden colours that are arranged or connected in the same order from head to tail, apple-green at the head end, through to herb-green at the tail end. In both cases, the colours provide regional territories, leading to structures with distinct identities developing along the head-tail axis. However, the way that this common map is interpreted and eventually becomes manifest in the visible features of the organism is quite different in each case. In insects, it becomes manifest as a distinction between the various types of segments of the animal whereas in vertebrates it affects different structures, most notably the types of bones arranged along the backbone. Anatomy therefore reveals this underlying map of hidden colours only indirectly: through the way the map is interpreted and eventually becomes manifest in the animal.

It follows that even if animals have a similar set of underlying colours, this unity might only be dimly perceived at the level of their anatomy. That is why many of Geoffroy's detailed correspondences, such as between vertebrate ribs and insect legs, proved to be unconvincing: he was trying to establish a unity of plan based purely on anatomical features. In this sense Cuvier was right to disparage some of Geoffroy's comparisons. Yet although we might fault Geoffroy on some of his particular claims, his overall insight that there was a common underlying map that unified animals as different as insects and vertebrates did prove to be correct.

From an evolutionary point of view, these limitations of anatomical studies can be seen to depend on the degree of relatedness between the animals being compared. There is no difficulty, even from the most superficial inspection, in establishing the correspondence between a *human hand* and the hand of an *ape*. It is more difficult to see the relationship between a human arm and a *horse's leg* or a *whale's flipper*. In these cases, if we only look at them from the outside, we might conclude that they are entirely different types Of appendage. The correspondence is greatly clarified, however, by looking at the arrangement of bones in the skeleton. What looks superficially different is seen to reflect a common arrangement of bones. But when we get to more distantly related animals, such as insects and vertebrates, even anatomical comparisons become of limited use. Based on these, we might have reasonably concluded that the basic layout of these animals had nothing in common (unless, like Geoffroy, we believed in a fundamental unity of plan). The anatomical layouts could have evolved completely independently of each other. To get a deeper insight, we need to look at the map of hidden colours that underlies the anatomy of the organism. Then we see that there is a unity even though it is interpreted and becomes manifest in very different ways in the anatomy of the animal. The map provides a hidden skeleton, a set of underlying connections that does allow meaningful comparisons between very diverse organisms to be made.

The green hidden colours provide an absolute or immutable map. There have been alterations in the *homeobox genes* needed for the green colours during evolution, such as extra duplications of individual *homeobox genes*. Even entire dusters of homeobox genes have been duplicated to give the four dusters present in mice and humans of today. These alterations may have allowed the map of green colours to have been modified to some extent. Nevertheless, the basic order of green hidden colours from head to tail, the connections in the map, does not seem to have changed in a fundamental way during the six hundred million years that separate insects and vertebrates from their common ancestor.

What role might the *homeobox genes* have had in this common ancestor of insects and vertebrates? It seems most likely that, as with organisms alive today, its *homeobox genes* would also have coded for a set of hidden colours that gave distinctions from head to tail. In this case, however, thc various territories of colour may not have been interpreted to give distinctive segments, like those of flies, nor different type of vertebrae, as in mice. Rather, they could have been manifested in a different way again.

A good illustration comes from the study of *homeobox genes* in the nematode *Caenorhabditis elegans*, a tiny worm-shaped animal (about one millimeter long) that belongs to a quite distinct group (phylum) from insects or vertebrates. This worm has neither segments nor vertebrae, yet it still has a duster of *homeobox genes* (the worm has only a set of four genes in its cluster as compared to the eight or more observed in insects and vertebrates). As with the other animals Each worm homeobox gene is expressed in a distinct region of the animal, producing territories of green hidden colour from head to tail. The hidden colours are then interpreted by genes so that different parts of the worm assume distinct identities along the head-tail axis. In this case, though, the identities do not refer to segments or vertebrae, but to groups of cells that are repeated along the worm's length. These cell groups are not surrounded by a hard outer casing, like insect segments, nor do they produce bony regions like vertebrae; they simply form characteristic regions of the worm. In other words, the hidden colours still provide a map or frame of reference from head to tail, but this is interpreted and eventually becomes manifest differently for the worm than for vertebrates or insects.

We can conclude that the common ancestor of insects and vertebrates (and nematode worms) most likely had a cluster of *homeobox genes* that provided distinctive identities along its head-tail axis. However, the way this hidden map was interpreted and became manifest in this early animal was probably quite different from what we see in humans or flies today. There is a remarkable unity in the map of some hidden colours between animals, which has been preserved for hundreds of millions of years; yet the way this becomes manifest in their anatomy can be very different.

6

EVOLUTION AND CREATIONISM

The reappearance of evolution in biology courses proved to be a stimulus for creationists, and their voices and activities increased. At first, only a few creationists were actively involved outside their community, but they proved to be skillful, determined, and effective. For example, the efforts largely of one couple in Texas were sufficient to make the adoption of biology books that discussed evolution extremely difficult in that state.

The major voices for creationism were those of ten men with advanced university degrees who in 1963 formed the *Creation Research Society* and later, in 1972, founded the *Institute for Creation Research*, an educational institution with faculty, students, and research programs. A pamphlet from the *Creation Research Society*—issued at the society's own creation in 1963—gave a brief history of the movement, a list of activities, and requirements for membership. The credo of the organization read:

1. The *Bible* is the written Word of *God*, and because it is inspired throughout, all its assertions are historically and scientifically true in all the original autographs. To the student of nature, this means that the account of origins in *Genesis* is a factual presentation of simple historical truths.
2. All the basic types of living things, including man, were made by direct creative acts of *God* during the creation week described in *Genesis*. Whatever biological changes have occurred since creation week have accomplished changes only within the original created kinds.

3. The great flood described in *Genesis*, commonly referred to as the *Noachian flood*, was a historic event worldwide in its extent and effect.

Two members of the institute, Henry M. Morris (now retired) and Duane Gish, were until recently the principal professional creationists in the United States. They, along with other members of the institute, set the pattern for the activities of the antievolutionists since the 1960s by promoting what they claim is an alternative to the scientific evidence for evolution, namely, *creation science*. Creation scientists refuse to accept merely on faith that the story of creation in *Genesis* is true and to ignore the scientific evidence for evolution. Instead, they try to discredit the scientific evidence for evolution and to assemble their own body of scientific evidence to support the P version of creation in Genesis.

Creation Scientists' Approach

The "*bible*" of the creation science movement is Scientific Creationism, edited by Morris and first published in 1974. It appeared in two editions, a general edition and one for public schools. The general edition discusses the evidence used by evolutionists and suggests that none of it proves that evolution occurred. The concluding chapter of the general edition attempts to present data that prove divine creation and are in full accord with a literal interpretation of the *Bible*. The public school edition "*deals* with all the important aspects of the creationist-evolution question from a strictly scientific point of view, attempting to evaluate the physical evidence from the relevant scientific fields without reference to the *Bible* or other religious literature. It demonstrates that the real evidences dealing with origins and ancient history support creationism rather than evolutionism".

The major ideas developed in creation science deny the general conclusions of biologists and geologists. For example, the Earth is assumed to be very young, 10,000 years being the outside limit. The authors reject radiometric dating by arguing that no one was present to see when the strata of the geological column were laid down. Since there can be no direct evidence of age, any such estimates must be uncertain at best. We can be certain that no evolutionist observed events about 4.5 billion years ago, but come to think of it, was there any creationist on hand 10,000 years ago to record what happened? Many things in science, especially historical subjects such as evolution, can be studied only by seeking indirect evidence of past events—the detective's approach. The utility of radiometric dating has long been accepted by scientists, especially when two different methods give essentially identical results. There was no way that the formation of ancient strata could have been seen by geologists.

An important implication of evolution is that there must have been intermediates, or transitional forms, between major groups if members of one group evolved into another. Many examples of intermediates are now known, and thus many "*missing links*" are in fact no longer missing. *Scientific Creationism* argues these new discoveries away with a surprising analysis. It notes that *Archaeopteryx* has both feathers and teeth and that feathers are characteristic of birds and teeth of reptiles. One paleontologist is quoted who classifies *Archaeopteryx* as a bird because it has feathers. The creationists conclude, "Thus, *Archaeopteryx* is a bird, not a reptile-bird transition. It is an extinct bird that has teeth. Most birds don't have teeth, but there is no reason why the Creator could not have created some birds with teeth". This is a fascinating solution, but it is not based on an understanding of how the biological system of classification works. The system of classification has no place for intermediates. *Archaeopteryx* could also have been classified as a reptile—as some fossils of it actually were. Creationists could have argued just as easily that *Archaeopteryx* was an extinct reptile with feathers and that there is no reason why the Creator could not have created some reptiles with feathers.

An important feature of creationism is the belief that the *Noachian flood* was a worldwide phenomenon that submerged the highest mountains for a period of months. All life was destroyed except the few individuals of every species that survived with Noah and his family in the Ark. This flood was responsible for depositing vast amounts of silt that became the layers of sedimentary rocks, with their fossil organisms. The creationists' date for this cataclysm is about 2350 b.c..

Another creationist, Austin (1994), has proposed that the Grand Canyon of the Colorado River was formed a few thousand years ago during Noah's flood. The data and arguments he uses are not those accepted by professional geologists, and a detailed response to Austin's theory has been provided by Elders (1998). Yet the familiar problem remains. To a reader with little or no knowledge of geology and of what counts as evidence in science, Austin seems to make a strong case. Professional geologists, however, agree that there is no geological evidence whatsoever for a worldwide flood that covered the highest mountains and destroyed all life less than 5,000 years ago.

Scientific Creationism ends with the following statement: There seems to be no possible way to avoid the conclusions that, if the *Bible* and *Christianity* are true at all, the geological ages must be rejected altogether. In their place, as the proper means of understanding earth history as recorded in the fossil-bearing sedimentary rocks of the earth's crust, the great worldwide Flood so clearly described in Scripture must be accepted as the basic mechanism. The detailed correlation of the intricate geophysical structure of the earth with the true Biblical framework of history will, no doubt, require a tremendous amount of research and study by Bible-believing scientists. Nevertheless, this research is urgently needed today in view of the world's increasing opposition to the *Biblical Christian faith*. The vast complex of godless movements spawned by the pervasive and powerful system of evolutionary uniformitarianism can only be turned back if their foundation can be destroyed, and this requires the re-establishment of special creation, on a Biblical and scientific basis, as the true foundation of knowledge and practice in every field. This therefore must be the primary emphasis in Christian schools, in Christian churches and in all kinds of institutions everywhere. It is hoped that this book will provide the information necessary to undergird and energize the movement.

Scientific Creationism seems to have done little, in fact, to "*undergird and energize*" the "*Bible-believing scientists*" to discover the data that will throw out what biologists and paleontologists have learned during the last few centuries about Earth's history. Expeditions to Mount Ararat have found there something that suggests the remains of Noah's Ark, but the evidence is not conclusive. Other creationists visit sites where geologists have found that the forces of mountain building have not only pushed up layers of rocks but turned them over. Geologists rely on study of the fossils in the strata to explain what happened, namely, that the strata have been reversed and, hence, the oldest are on the top, not the bottom. Creationists use these upside-down strata to argue that the whole story of successive layers, each representing a capsule of time, is wrong.

One of the most interesting attempts to refute the notion of the immensity of geological time and the unique fossils in the strata is the creationists' claim that human footprints and dinosaur footprints occur in the same geological strata. This would mean, of course, that dinosaurs and human beings *coexisted*—a notion that is definitely not compatible with the history of life as understood by biologists and paleontologists. Current estimates place the extinction of dinosaurs at about 65 million years before the appearance of human beings. Those controversial footprints are in the Cretaceous rocks along the banks of the Paluxy River in Texas. Some have been identified by geologists as indeed caused by dinosaurs. Other depressions are called *human footprints* by some local people including religious leaders. These have been examined extensively by geologists, and it is now accepted that the "*human footprints*" have a variety of origins. Some are badly eroded dinosaur footprints. Others have been carved by local citizens and hence are the result of human hands rather than human feet.

Some local people recount that in the 1930s stoneworkers made a business of carving dinosaur and human footprints in the same slab and selling them to devout visitors who wished clear evidence for the correctness of Genesis and the denial of Darwin. After many years of debate and careful study by geologists, the better-informed creationists have admitted that there are no human footprints in the Cretaceous strata along the banks of the Paluxy. This myth continues, however, in the minds of many rank-and-file creationists.

Another creationist argument against evolution is that it violates the *second law of thermodynamics*. This fundamental law of the physical sciences holds that matter achieves and maintains complexity only with a constant supply of energy. If animals do not have a constant input of energy from food, they die and their complex structure decays to relatively simple molecules. Green plants are able to use the energy of sunlight for the synthesis of their living substance. Animals use the products of the plant world to produce their own structures and the energy to maintain those structures. If an animal was in a closed system, that is, with no supply of external energy, the existing energy in the molecules that comprise its body would slowly be lost and death ensure. Some creationists argue that since evolution entails an increase in the *complexity of organisms*—from simple one-celled creatures to whales, elephants, redwood trees, and human beings—it violates the *second law*. Thus there can be no evolution from simplicity to complexity. Creationists seem to forget that they, along with nearly all complex animals, began life as a single cell, which developed into a complex adult—an example of a level of complexity the second law fails to prevent. Why? This is possible because organisms do not live in a closed environment—food is required to sustain lives.

Evolution per se does not require additional energy—it involves only individuals being born, growing up, reproducing, and dying.

Organisms use energy whether or not they are evolving. They evolve only if the proportions of different genetic types in the offspring differ from those in the parental generations. Not one of the prominent creationists believes his argument about the second law of thermodynamics, but reference to it can have a powerful effect on a naive audience. So if it works, why not use it?

The creationists have not been able to present any evidence supporting their arguments for a young Earth, a worldwide flood, or an interpretation of the paleontological data that would support the accuracy of the P version in *Genesis*. They have, however, been very successful in presenting their case to the public. In recent years Gish and other creationists have lectured widely to church groups and to college and university communities both in the United States and abroad. Until the 1990s, the lectures followed a standard pattern and dealt with just a few topics. The footprints on the Paluxy were presented repeatedly as was the second law of thermodynamics. Both in lectures and in published works such as *Scientific Creationism* acceptance of the creationists' arguments depends on an inadequate understanding of science or firm religious beliefs.

Scientific Creationism is well-written and is far less demanding than a scientific account of evolution. For a poorly informed person it would probably be more appealing than a scientific book on evolution. The creationists' story is easy to understand and may support religious beliefs that have been held since childhood. The evolutionists' story demands a more open mind and a good background in science—or else a simple faith that scientists are presenting their findings in a fair and honest manner.

Professional creationists spend a great deal of their time on speaking tours, and some of them are quite up-to-date on recent research in evolutionary biology and paleontology. Thus they know where the gaps are in the data for evolution. The goal of the creationists is to emphasize the gaps and sow doubt in the minds of the audience about the adequacy of the evidence for evolution. They might pose a question like: "Evolutionists claim that whales evolved from land-living mammals. Where is the evidence?"

Until recently there was none. The scientists' answer is that there is considerable evidence from the fossil record that traces the evolution of some species, but one cannot expect to have fossil evidence for the origins of all living creatures. Most fossil records are and probably always will be incomplete. A few well-documented cases are sufficient, and this number increases yearly. For example, just a few years ago fossils that indicate an intermediate stage between a terrestrial mammal and a whale were discovered in Pakistan.

Using this skeptical approach, a skilled creationist can easily best an evolutionist in debate. The audience can be made to wonder why there is no evidence to answer the creationist's specific question. A poorly informed audience is not always prepared to accept that if there are a few cases showing the evolution of a group of species—the horses, for example—that is enough.

In a sense this is a matter of the public's not understanding how science works and what is accepted as proof. No science is ever complete; its statements represent only the best analysis at the time, based on the best available evidence. Consider the case of medicine: Although the medical profession knows a great deal about many diseases, everything is not known about any one disease. The procedures of medicine represent the best that can be done with what is known. Nevertheless, a sick person is unlikely to refuse the services of a physician by maintaining, "I will not accept medical treatment for heart disease until the profession knows enough to cure the common cold."

Picking away at the gaps in evolution is a reasonable and effective approach if the purpose is to destroy that point of view and imply that the alternative, divine creation, must be correct by default. The goal of such a strategy is, most certainly, not to provide an alternative rational explanation for the diversity of living animals and plants. Quite often the biologist or paleontologist who debates a creationist fails to realize that the creationist's purpose is not to inform but to win the battle for a particular religious point of view. The episode is more like a court trial, where each lawyer's goal is to win for the person he or she represents. If the victory for the client is also a victory for justice, so much the better, but justice is not the lawyer's primary concern.

In spite of the careful way creationists might present their case, in the final analysis it is based on belief in the story of divine creation told in the P version of *Genesis*, not on scientific evidence. The creationists' approach is to claim that evolution cannot be correct because there are too many unknowns, and therefore to conclude that the *Genesis* story must be true by default. But even if evolution could be falsified, why should the default position be *Genesis*? What about the many different accounts of creation that are parts of the sacred traditions of other religions? What is the basis for excluding them? Creationists have no logical answer for that question.

Most individuals who have informed themselves deeply about the data agree that evolution provides a much better account for the changes in the history of life over time and for the diversity of life we see today than does creation science, which is nothing more than old-fashioned creationism—without any science. The name creation science is regarded by many noncreationists as an oxymoron because creation implies a supernatural force, and science can never use supernatural forces to explain its data.

Biologists and paleontologists have assembled information fully compatible with the concept of evolution as guided by natural processes, whereas none of the few types of data that creationists offer in support of divine creation pass muster in the sciences. Much remains to be discovered about evolution, but those who study these problems and are involved in new discoveries accept the overarching concept of evolution as true beyond all reasonable doubt. Most antievolutionists around today are people who are unfamiliar with these data and with the procedures used to gain scientific understanding, or who choose to reject the theory of evolution for religious or other reasons.

New Creationism

In the 1990s, the style and substance of the professional creationists' attacks on evolution changed somewhat in response to the times. The leaders of the movement downplayed or failed to accept the standard antievolution arguments such as the supposed incompatibility of evolution with the second law of thermodynamics, the supposed lack of a convincing number of fossils that are intermediate between major taxonomic groups of organisms, and the reputed footprints of human beings alongside those of dinosaurs in the geological strata along the banks of the Palusky River. The errors of these arguments came to be accepted by many creationists in the last decade.

The cutting edge of the creationist movement is now moving beyond the traditional literal interpretation of *Genesis*. The position that creation as described in the P version of *Genesis* is right and so evolution must be wrong is now being replaced by the argument from design. This notion is that some aspects of life and living species are so complicated and complex that they cannot be explained by science. There must have been an Intelligent Designer to have been able to create something as complex as the human eye, for example.

The intelligent design movement is a slight variant on the position of the natural theologians of the nineteenth century and before. It was one that appealed to many naturalists and others such as Charles Darwin and Thomas Henry Huxley. One of its most prominent advocates was the theologian William Paley, who in 1802 argued that some of the structures of animals are so beautifully and effectively adapted to their environment that they could not have "*just happened.*" There must have been a designer, namely, God. Paley's analysis, appearing when little was yet known about living organisms, made a powerful impression in England in the nineteenth century. It was an argument more acceptable to many devout Christians than a belief that creation was accomplished in only six days and as recently as 4000 b.c.

This is Paley's famous opening of his Natural Theology: In crossing a heath, suppose I pitched my foot against a stone, and were asked how the stone came to be there; I might possibly answer, that, for any thing I knew to the contrary, it had lain there for ever: nor would it perhaps be very easy to show the absurdity of this answer. But suppose I had found a watch upon the ground, and it should be inquired how the watch happened to be in that place; I should hardly think of the answer which I had before given, that, for any thing I knew, the watch might have always been there. Yet why should not this answer serve for the watch as well as for the stone? Why is it not as admissible in the second case, as in the first? For this reason, and for no other, viz. that, when we come to inspect the watch, we perceive (what we could not discover in the stone) that its several parts are framed and put together for a purpose, e.g. that they are so formed and adjusted as to produce motion, and that motion so regulated as to point out the hour of the day.

Paley described many of the anatomical and other characteristics of living organisms and argued that just as there is a watchmaker who makes a watch for the purpose of telling time there must have been the equivalent of a divine watchmaker for designing the wonderful structures of the species of living creatures. That was, of course, God. Paley concluded:

I shall not, I believe, be contradicted when I say, that, if one train of thinking be more desirable than another, it is that which regards the phenomena of nature with a constant reference to a supreme intelligent Author. . . . The world thenceforth becomes a *temple*, and life itself one continued act of *adoration*. The change is no less than this, that, whereas formerly *God* was seldom in our thoughts, we can now scarcely look upon any thing without perceiving its relation to him. Every organized natural body, in the provisions which it contains for its sustentation and propagation, testifies a care on the part of the Creator, especially directed to these purposes. . . . The works of nature want only to be contemplated. When contemplated, they have every thing in them which can astonish by their greatness: for, of the vast scale of operation, through which our discoveries carry us, at one end we see an intelligent *Power* arranging planetary systems, fixing, for instance, the trajectory of *Saturn*, or constructing a ring of two hundred thousand miles diameter, to surround his body, and be suspended like a magnificent arch over the heads of his inhabitants; and, at the other, bending a hooked tooth, concerned and providing an appropriate mechanism, for the clasping and reclasping of the filaments of the feather of the humming-bird. We have proof, not only of both these works proceeding from an intelligent agent, but of their proceeding from the same agent. . . . Therefore one mind hath planned, or at least hath prescribed, a general plan for all these productions. One Being has been concerned in all.

Under this stupendous Being we live. Our happiness, our existence, is in his hands. All we expect must come from him. Nor ought we to feel our situation insecure. In every nature, and in every portion of nature, which we can descry, we find attention bestowed upon even the minutest parts. The hinges in the wings of an earwig, and the joints of its antennae, are as highly wrought, as if the Creator had nothing else to finish.

Thus, a watch fulfills its purpose because the watchmaker designed it to do so, and the species of animals and plants live their specialized lives because of the ways the Supreme Author designed them. At the time Paley was writing, little was known of the details of how our bodies function. It seemed impossible to understand how a structure as complex as the human eye might function or to accept that it "just happened."

Of course, no evolutionist today would claim that eyes "*just happen*." The working hypothesis is that they are the consequence of the selection of gene mutations that initially produced small areas in the skin that were light-sensitive in very simple creatures. Slowly, over great expanses of time and in a succession of species, the light-sensitive areas are thought to have become larger and more complex. After many millions of years they had become the vertebrate eyes. At all stages, it is assumed mutations were selected that made the light-sensitive areas more effective. In other words, being able to detect light was highly adaptive for some types of animals.

There is no sequence of fossils that shows this proposed evolution of the vertebrate eye. The soft parts of organisms are almost never fossilized, and in situations such as this the best that can be done is to search for living animals with light-sensitive structures of varying levels of complexity. It is now known that there exists a sequence of living animals, with a graded series of ever more complex light-receiving structures, that shows what the evolutionary story might have been. The story is far from complete, but one no longer need invoke super-natural processes to explain the eye—as Paley was forced to do.

Natural theology provided an alternative to *biblical theology* in explaining the natural world. There was still a causal agent, the God of *Genesis* or the *Intelligent Designer* of the natural theologians. *God* remained in His firmament with the natural theologians seeking to know Him by studying nature and the theologians seeking knowledge through biblical analysis. Both groups continued to rely on a supernatural force. The change in thought patterns that began in the Enlightenment made an increasing number of individuals restive with explanations of natural phenomena that invoked the supernatural. Studies of phenomena in astronomy, physics, and chemistry were increasingly able to reject the supernatural and to provide explanations that involved only natural identifiable things and processes.

In the middle of the nineteenth century Charles Darwin was able to add biology to those other sciences able to ignore supernatural explanations. Nevertheless creationists have continued to exist, and the intelligent design advocates are now active and effective with the general public, effective in part because the scientific community tends to ignore them completely.

The intelligent design program is rejected by scientists and others because I.D. creationists believe that some things about organisms are so complex that they are not just unknown but ultimately unknowable and, hence, are the work of an *Intelligent Designer*. This point of view was perfectly valid when there was essentially no knowledge of the physiology of cells on which life depends. How, for example, could one ever understand what must be an incredibly complex series of chemical reactions that occur with great speed in the cells of animals and plants that were so small as to be almost invisible? In the years since World War II, with the aid of a battery of new techniques, the enormous complexity of cell metabolism has been documented. Today there seems to be no aspect of the life of a cell that is unknowable, a situation similar to geologists gaining the ability to determine geological time. At the end of the nineteenth century it seemed that there could never be a method that would give confirmable answers. The discovery of radioactivity eventually provided a series of acceptable methods for geological clocks that tell accurate time and, most important, agreed with one another as to what the time is.

Fundamental to the commitment of any working scientist is the position that what we do not know today we may know tomorrow. The advance of science so far has shown this approach to be highly successful, and scientific knowledge continues to expand at an ever-increasing rate. Perhaps some things will never be known, especially events in the distance past for which no direct or indirect evidence exists. But even if we never know, for example, the mating habits of dinosaurs, that lack of concrete knowledge does not add up to an argument that dinosaurs did not mate and evolve. Recent books by Phillip Johnson (1991, 1997) and Michael Behe (1996) give more information about the intelligent design position. A balanced analysis is provided by Pennock (1999).

What can be the explanation for this continuing argument of creationists and evolutionists? Two social scientists, Francis B. Harrold and Raymond A. Eve (1995), have studied the creationist movement for years and have this to say:

Average Americans are ill-equipped to evaluate the claims of scientific creationism. They can hardly be expected to analyze creationist claims about the second law of *thermodynamics*, for instance, if they lack any idea of what the first or third laws are about. If they have no notion of the wealth and range of evidence for human evolution, they may find perfectly reasonable the claim that all Homo erectus fossils are either apes or modern humans. If they subscribe to the common misconception that evolutionary change is the result of "*blind chance*," they may perceive creationist arguments of its extreme improbability as compelling. If they have little understanding of the roles of critical thinking and "*rules of evidence*" in science, they may cling to unsubstantiated beliefs even after seeing them effectively refuted; even if they do change their minds, they may revert after a time to their former beliefs. The efforts of scientific creationists have not affected the scientific consensus on evolution because creation scientists do not function in the research tradition of modern science. . . . They launch scattershot attacks on the evolutionary consensus based on the notion that evidence that damages evolution is evidence that supports creationism. Their actual criticisms of scientific finds are characterized by misunderstandings, omissions, and distortions. They do not attempt to develop detailed alternative scientific explanations of the data they claim to have "*unexplained*" in their criticisms of the consensus. Instead, they appeal to the supernatural. . . . It seems safe to conclude that creationist rhetoric has indeed bolstered the ideology's credibility . . . [Their] primary interest is not so much to develop an alternative science as to defend a traditional worldview or even advocate the establishment of a conservative Christian theocracy. In either of these cases, there would be no role left for science to play, except when its findings could be used to support scripture. Modern science, generally, which rationally deduces hypotheses from existing knowledge, tests them with data, and then lets the philosophical and theological chips fall where they may, would have little place in a society such as the creationists envision.

Back to the Courts

From 1923 to 1928, Arkansas, Mississippi, Oklahoma, and Tennessee passed laws forbidding the teaching of evolution. In other states such efforts were defeated. In 1968 the Supreme Court of the United States, in Epperson v. Arkansas, struck down the Arkansas statute on the grounds that it was in conflict with the establishment clause of the First Amendment of the Constitution of the United States, which reads as follows: "Congress shall make no law respecting an establishment of religion." The court interpreted creationism as based on religion and as promoting the views of a particular sect—the Judeo-Christian tradition.

Having failed in the courts to ban evolution from the classroom, the creationists tried a different approach. If it was impossible to keep evolution out of the schools, then an attempt should be made to get creationism in. They argued that evolution was "*just a theory*," using the term theory to signify an iffy hypothesis. After all, no scientist ever claimed to have seen one species evolving into another, so creationists argued that evolution could hardly have been considered a fact. *Creationism* as derived from the P version in *Genesis* could not be proven either, so both were just different ways of looking at nature. A just solution, therefore, would be to present both views and let the students decide. This was the equal-time position.

Scientists indeed cannot object to creationists saying that evolution is "*just a theory*," because scientists themselves use the word theory quite readily to describe evolution and other major natural phenomena, such as the *germ theory of disease*, the theory of gravitation, cell theory, or *genetic theory*. In common usage, the word theory usually means a questionable idea—a hypothesis as yet unproven. This is the standard way the word is used in creationists' speeches and in much of public discourse. It is also the way legislatures and school boards use the term when they seek to banish evolution from the schools. But scientists use the word theory to mean a large, overarching concept that organizes a vast body of data.

Wendell Bird, a lawyer then on the staff of the Institute for *Creation Research*, prepared a draft statement advocating equal time for the "theories" of evolution and of creation as described in Genesis. This draft was used by Paul Ellwanger, a therapist and then head of a creationist group called Citizens for Fairness in Education, to prepare a model bill for state legislatures to consider. It was the basis of bills introduced in about two dozen state legislatures. Only in Arkansas and Louisiana were bills passed, and both were challenged in the courts.

So for the second time in just a few years Arkansas became a focal point in the evolution-versus-creationism debate. *Arkansas' law*, called the *Balanced Treatment* for Creation-Science and Evolution-Science Act (Act 590), was signed by the governor on March 19, 1981. It demanded that equal time be given to the teaching of evolution and creationism. On May 27 of that year a suit, McLean v. Arkansas Board of Education, was filed challenging the constitutional validity of Act 590 on the grounds that it violated the establishment clause of the Constitution. Judge William R. Overton, of the Arkansas Supreme Court, heard the case.

The plaintiffs, who wished Act 590 to be declared unconstitutional, assembled an impressive group of scientists to discuss the nature of science and of creation science. Among them were Stephen Jay Gould of Harvard University and William V. Mayer, Director of the Biological Sciences Curriculum Study. There were also professional philosophers and religious leaders among the witnesses for the plaintiffs. The defendants—those favouring equal time—were unable to produce any well-respected scientists to support their arguments. They never called upon either Morris or Gish of the Institute for Creation Research, out of fear that the court would probably realize that their "*creation science*" was not based on scientific evidence at all. This was the best opportunity yet for the creationists to present their case, but they could develop no more than a feeble position.

Judge Overton rendered his judgment on January 5, 1982, and concluded with the following analysis:

In the 1960's and early 1970's, several Fundamentalist organizations were formed to promote the idea that the Book of Genesis was supported by scientific data. The terms "*creation science*" and "*scientific creationism*" have been adopted by these Fundamentalists as descriptive of their study of creation and the origins of man. . . . The fact that creation science is inspired by the Book of Genesis and that Section 4(a) [of Act 590] is consistent with a literal interpretation of Genesis leave no doubt that a major effect of the Act is the advancement of particular religious beliefs. . . . The conclusion that creation science has no scientific merit or educational value as science has legal significance in light of the Court's previous conclusion that creation science has, as one major effect, the advancement of religion.

The essential characteristics of science are: (1) It is guided by natural law; (2) It has to be explanatory by reference to natural law; (3) It is testable against the empirical world; (4) Its conclusions are tentative, i.e., are not necessarily the final word; and (5) It is falsifiable. Creation science . . . fails to meet these essential characteristics. Implementation of Act 590 will have serious and untoward consequences for students, particularly those planning to attend college.

Evolution is the cornerstone of modern biology, and many courses in public schools contain subject matter relating to such varied topics as the age of the earth, geology, and relationships among living things. Any student who is deprived of instruction as to the prevailing scientific thought on these topics will be denied a significant part of science education. Such a deprivation through the high school level would undoubtedly have an impact upon the quality of education in the State's colleges and universities, especially including the pre-professional and professional programs in the health sciences.

Judge Overton had agreed with the plaintiffs that Act 590 was unconstitutional. This was an astonishing victory for the teaching of evolution, and it had the effect of abolishing efforts in other states to teach creationism along with evolution, for a few years at least.

In 1987 the United States Supreme Court, in Edwards v. Aguillard, struck down a Louisiana act that required equal time for the teaching of *evolution* and *creationism*. Once again, it was judged that the basis for the law specifying that creationism be taught was religious belief, not science, so the law was in conflict with the establishment clause. The majority opinion of the court was written by Justice William

Brennan. One of his key statements was: The preeminent purpose of the Louisiana legislature was clearly to advance the religious viewpoint that a supernatural being created human kind. The term "creation science" was defined as embracing this particular religious doctrine by those responsible for the passage of the Creationism Act. . . . The legislative history therefore reveals that the term "creation science" as contemplated by the legislature that adopted this Act, embodies the religious belief that a supernatural creator was responsible for the creation of human kind.

and, hence, this is prohibited by the establishment clause.

The following year, in 1988, the attorney general of Tennessee was asked if a teacher in a public school was free to teach all theories of the origin of life for the purpose of enhancing the effectiveness of science instruction. His answer was no, if the theory had a religious basis. This conclusion was based on various legal opinions, especially the then-recent Edwards v. Aguillard case in Louisiana. The attorney general made it clear, however, that biblical accounts of creation could be included in courses in history, ethics, or comparative religion.

Neither Judge Overton's ruling in the 1982 Arkansas case, nor that of the United States Supreme Court in the 1987 Louisiana case, nor any of a handful of others—all dismissing the teaching of creationism as science—ended the creationists' efforts to curb or mitigate the teaching of evolution. The contest shifted from state legislatures to departments of education, school boards (Kansas being a notable example in 1999), and individual schools. Among the many creationist-inspired events that have taken place recently are the introduction of antievolution bills in state legislatures (none has passed), actions taken by local school boards to restrict or abolish the teaching of evolution, textbooks failing to win adoption because they give too much attention to evolution, the active teaching of creationism in many schools in spite of the decisions that this is in conflict with the Constitution, the requirement of disclaimers in textbooks such as "*evolution is just a theory*," state science curriculum standards omitting evolution, and an especially innovative step—the gluing together of the offending pages discussing evolution in the textbooks used by students. This last strategy may in fact guarantee that teenagers will become tremendously eager to learn everything they can about evolution—or anything else that adults glue shut.

Legal opinions—all the way up to the United States Supreme Court—prohibiting the teaching of creationism have not brought peace to the classrooms. The most destructive effect of the evolution-versus-creationism controversy remains the chilling effect it has had on biology teachers. Although science has prevailed in major legal challenges to the point that there are currently no laws on the books in any state prohibiting the teaching of evolution or permitting equal time for creationism, the problems for men and women trying to teach evolution are probably as difficult in some communities today as they have ever been.

Evolutionists Fight Back

The increased activity of professional creationists and the spread of their views beginning in the 1960s brought forth a major reaction from religious, educational, and scientific organizations. Scientists reacted promptly, and statements against the teaching of creationism or against giving it equal time with evolution were made by the National Academy of Sciences, American Association for the Advancement of Science, American Anthropological Association, Association of Physical Anthropologists, American Astronomical Society, American Chemical Society, American Geophysical Union, American Institute of Biological Sciences, American Physical Society, American Psychological Association, American Society of Biological Chemists, American Society of Parasitologists, Geological Society of America, Society for the Study of Evolution, Society of Vertebrate Society, and many state academies of science and organizations of science teachers.

A typical statement was made in 1972 by the Commission on Science Education of the American Association for the Advancement of Science, the world's largest scientific society:

The Commission on Science Education of the American Association for the Advancement of Science is vigorously opposed to attempts by some boards of education and other groups to require that religious accounts of creation be taught in science classes.

During the past century and a half, the earth's crust and the fossils preserved in it have been intensively studied by geologists and paleontologists. Biologists have intensively studied the origin, structure, physiology, and genetics of living organisms. The conclusions of these studies are that the living species of animals and plants have evolved from different species that lived in the past. The scientists involved in these studies have built up the body of knowledge known as the biological theory of the origin and evolution of life. There is no currently acceptable alternative scientific theory to explain the phenomena.

The various accounts of creation that are part of the religious heritage of many people are not scientific statements or theories. They are statements that one may choose to believe, but if he does, this is a matter of faith, because such statements are not subject to study or verification by the procedures of science. A scientific statement must be capable of test by observation and experiment. It is acceptable only if, after repeated testing, it is found to account satisfactorily for the phenomena to which it is applied.

Thus the statements about creation that are part of many religions have no place in the domain of science and should not be regarded as reasonable alternatives to scientific explanations for the origin and evolution of life.

The prestigious National Academy of Sciences issued in 1984 a lengthy and authoritative statement, Science and Creationism: A View from the National Academy of Sciences, in support of evolution and in opposition to the teaching of creationism. A second edition was released early in 1999. The academy has also produced a book especially designed for schoolteachers, Teaching about Evolution and the Nature of Science (1998).

Many religious groups have also issued statements against the teaching of creationism in schools. The General Assembly of the United Presbyterian Church in the United States developed this position in 1982 from which the following is extracted:

Affirms that, despite efforts to establish "*creationism*" or "*creation-science*" as a valid science, it is teaching based upon a particular religious dogma as agreed by the court.

Affirms that, the imposition of a fundamentalist viewpoint about the interpretation of *Biblical literature*—where every word is taken with uniform literalness and becomes an absolute authority on all matters, whether moral, religious, political, historical or scientific—is in conflict with the perspective on ***Biblical interpretation*** characteristically maintained by Biblical scholars and theological schools in the mainstream of Protestantism, Roman Catholicism and Judaism. Such scholars find that the scientific theory of evolution does not confiict with the interpretations of the origins of life found in Biblical literature.

Affirms that, exposure to the *Genesis* account is best sought through the teaching about religion, history, social studies and literature, provinces other than the discipline of natural science, and calls upon *Presbyterians*, and upon legislators and school board members, to resist all efforts to establish any requirement upon teachers and schools to teach "*creationism*" or "*creation science*." Among the other religious groups that have issued similar resolutions opposed to the teaching of creationism are the American Jewish Congress, the Central Conference of American Rabbis, the General Convention of the Episcopal Church, the Lexington Alliance of Religious Leaders, the Lutheran World Federation, the Roman Catholic Church, the Unitarian Universalist Association, and the United Methodist Church.

Not unexpectedly, many civil liberties organizations such as the American Civil Liberties Union, which was deeply involved in the Scopes trial, have issued statements. Part of a long ACLU position paper reads as follows:

Among the problems "*creation science*" creates in the academic environment is the foreclosure of scientific inquiry. The unifying principle of "*creationism*" is not the law of nature, but divinity. A divine explanation of natural data is not subject to experiment, it cannot be proved untrue, it cannot be disputed by any human means. Creationism necessarily rests on the unobservable; it can exist only in the ambiance of faith. Faith—belief that does not rest on logic or on evidence—has no role in scientific inquiry.

Vigilance requires firm and consistent opposition to every effort to use the nation's schools to teach any biblical text, including Genesis, as literal truth, either directly or disguised as "*alternative*" science. To reject creationism as science is to defend the most basic principles of academic integrity and religious liberty.

The position of the Catholic Church is of special interest when we remember its centuries-long opposition to scientific discoveries that provided evidence contrary to the official Church position. The Catholic Church has accepted evolution, the most definitive statement being made by Pope John Paul II to the Pontifical Academy of Sciences on October 22, 1996. The pope said that the theory of evolution is more than a hypothesis and that "it is indeed remarkable that this theory has been progressively accepted by researchers, following a series of discoveries in various fields of knowledge. The convergence, neither sought nor fabricated, of the results of work that was conducted independently is in itself a significant argument in favour of this theory". The question of the human soul must be looked upon in a slightly different manner, and John Paul II agrees with his predecessor, Pius XII, who in the encyclical Humani generis held that "if the human body takes its origin from pre-existent living matter, the spiritual soul is immediately created by *God*". John Paul then pointed out that although physical continuity—that is, the evolution of human beings—can be studied, the moment that the soul enters the body cannot. This assertion is not a problem for evolutionists because the soul cannot be identified or studied by the methods of science; it remains in the realm of religious beliefs. John Paul's statement is of great importance because it permits Roman Catholics to accept evolution. Nearly half of the world's two billion Christians are Catholics, and in the United States they are by far the largest denomination, totaling about 60 million.

Thus, as the twenty-first century begins there is a consensus among scientists, biblical scholars, most theologians of the mainstream religions, and educators that the theory of evolution provides a verifiable account for the origins of life and its changes over time, and that acceptance of evolution does not require a denial of one's religious beliefs.

7

Responding to the Environment

The German illustrator Ludwig Richter tells in his autobiography how as a young art student in the 1820s, he and three friends visited a famous beauty spot in Tivoli where they sat down to draw the landscape. All four of them were determined to draw as precisely as possible what was in front of them, not deviating in the slightest detail from nature: 'We fell in love with every blade of grass, every tiny twig, and refused to let anything escape us. Every one tried to render the motif as objectively as possible.' However, when they compared their efforts later on in the evening, they were surprised to see how different each picture was. Although the subject was the same, the pictures were as different as the personalities of the four artists.

The story shows that no matter how hard someone might try to imitate nature, the result is always an individual response of the artist rather than a carbon copy. The environment never determines or dictates what the picture will look like because so much depends on the way the individual artist reacts. Painting a landscape or portrait is an active process in which the artist responds to the surroundings rather than passively imitating them. Although the final picture may remind us of a scene, it is always through the eyes and reactions of the individual artist rather than an objective rendition of nature.

Now in describing the relationship between painting and the environment, we need to be careful to distinguish between two very different sorts of environment. First of all, there are the general surroundings of the artist, such as the lighting conditions, the landscape that is being painted, or the person who is sitting for a portrait. All of these conditions are what we commonly mean by the artist's external environment. A second aspect of the environment is the canvas or support on which the artist is working. This is a very particular part of the artist's surroundings, the part that he or she is actively working on.

The relationship of the artist to these two aspects of the environment—the canvas and the general surroundings—is quite different. The canvas is continually changing as a direct result of the artist's actions. Each time a brush stroke is applied, the artist modifies what is on the canvas, defining a new region of colour. The artist then reacts to this. Perhaps the colour is too strong, or it dashes with another colour on the canvas, or it fits in perfectly. Whatever the reaction, it will influence the next colours that are applied. These will in turn modify the canvas further, to produce yet more reactions in the artist. The artist is continually modifying as well as responding to the canvas, so the patterns change in a highly interactive and dynamic way. Although the canvas is physically external to the artist, outside his or her body, it is very much an integral part of the creative process, an inseparable part of the act of painting.

By contrast, the general surroundings of the artist, such as the landscape, model or lighting, are not modified to any great extent whilst the painting proceeds. The landscape may of course change as the sun sets or the clouds move, but these are not as a direct consequence of what the artist is doing. There may be the occasional bit of interference by the artist, such as asking a sitter for a portrait to turn slightly more this way or that, but this is relatively minor compared to the extensive interactions between the artist and the canvas. A movement or change in the subject may even be a source of irritation for an artist who might want everything to stay fixed; whereas changes in the canvas are an essential part of the painting process. So although both the general surroundings and the canvas are physically external to the artist, their relationship to the painting process is quite different. We might say that the overall surroundings provide an external environment whereas the canvas is an internal environment, an integral part of the creative activity itself.

We can get some curious effects when we confound these two aspects of the artist's environment. Look at Magritte's painting, called The *Human Condition*. It is a painting within a painting: a picture on an easel framed within a room and landscape. As the painting on the easel is exactly contiguous with the landscape behind, it seems to be both a part of the landscape and a representation of it. Magritte said of this picture: 'In front of a window, as seen from the interior of a room, I placed a picture that represented precisely the portion of landscape blotted out by the picture. For instance the tree represented in the picture displaced the tree situated behind it, outside the room. For the spectator it was simultaneously inside the room, in the picture; and outside, in the real landscape, in thought' Magritte's picture works by confounding the canvas with its general surroundings. However, the environment he shows, the room and window looking onto a landscape, is not truly external; it has been created by Magritte as part of the overall canvas. Although the surroundings of the easel may seem to be part of the external environment, they have actually been put there by the artist. The paradox arises because we confuse the external environment, the one that the artist does not greatly influence, with what is on the canvas itself. But no matter how realistic the painter's style, the general surroundings never become incorporated into the canvas. The canvas is not part of the external environment, it is internal to the act of painting, something that has been created through mutual interaction with the artist.

Developmental Responses

When we look at how an organisms development is related to its environment, we encounter a comparable situation to that in art. As we have seen, the growth and development of plants and animals are highly interactive processes in which each step is an interpretation of what went before. This is equivalent to the artist's internal dialogue with the canvas. Furthermore, as with painting, development does not occur in isolation but is always in the context of a particular set of environmental conditions. Factors such as temperature, humidity and light may all influence the way an organism develops. We shall see that it is not that these factors dictate how development will proceed, but rather that development responds in one way or another to them, just as the activity of an artist may respond to the external environment.

There are many different styles in which developing organisms respond to their environment, just as there are a variety of approaches to art. They range from cases in which the environment provides a rather uniform context, much as the general lighting conditions and ambience set the stage for an abstract painting, to cases in which variation in the environment plays an important role, nearer to what happens in representational art, such as landscape or portrait painting. Before getting into the mechanisms of how the internal systems of development are able to respond to the external environment, let me give a few examples of the various styles.

In many cases, the external environment is held relatively constant while the organism develops. The embryo is buffered from the outside world by, say, being protected within a mother's womb, or by being incubated in a nest by a parent. T is perhaps the type of his development we are most familiar with, in which the environment provides a relatively uniform context that allows the process to occur.

In other cases, development is more geared to *heterogeneity* in the environment. We already mentioned that some species of butterfly can develop in two different colour forms, say a spring and a summer variety, according to the environmental conditions at their time of development. This developmental response allows their colour to be adapted to different seasonal conditions, just as we vary our dress according to season.

Another example of how development may respond to the environment involves sex determination in some reptiles. Unlike ourselves, where sex depends on which chromosomes we inherit (females have two X chromosomes, males have one X and one Y), the sex of all crocodilians, most turtles and some lizards depends on the temperature at which they develop. Eggs incubated at low temperatures produce one sex, say a male, whereas those incubated at a higher temperature produce the other, a female. The survival of these species depends on variation in the environment: if the environmental temperature was always low or always high, all the offspring would be of the same sex, signalling the end of the species. In these cases, modifying development in response to heterogeneity in the environment is an essential feature of the system.

An even more intriguing form of environmental sex determination is in the sea worm *Bonellia*. The female of this species is about the size of a *plum* and has a proboscis about one metre in length. The male is a tiny worm-like animal a few millimeters long, and lives as a parasite within the uterus of the female. When the eggs fertilized by the male are liberated, they develop into free-swimming larvae. Any of the larvae that happen to settle on the proboscis of an adult female will develop into males, whereas those that remain free-swimming and gradually sink to the bottom become females. In other words, whether the larva develops into a large female or a tiny male depends on its external environment, whether it happens to land on an adult proboscis or not.

It is in plant development, however, that we see the most extensive modifications in response to environmental variation. Because they are rooted to the ground, plants continually have to modify their patterns of development according to local circumstance. A plant's direction of growth, branching pattern, leaf shape and flower production can all depend on the external environment. Because of this flexibility, plant development is often said to be plastic. In my view, this is misleading because plasticity conjures up the idea of a malleable substance whose shape and form are imposed from without. Plants, however, are not moulded by their external surroundings; they respond to them. For example, most shoots tend to grow towards light. This is not because the light draws or pulls the shoots towards it, but because plant shoots respond in a particular way to light, modifying their pattern of growth such that the tips bend towards the source of illumination. Roots will typically react in a different way, tending to grow away from light.

Similarly, plants can respond in various ways to the direction of gravity. Even when grown in complete darkness, shoots tend to grow upwards, in the opposite direction to the force of gravity, whereas roots respond in the opposite way by growing downwards. It is not that the roots are pulled down by gravity; they change their orientation of growth in response to it.

One of the most striking environmental responses is displayed by some amphibious plants which grow partly submerged in water. The leaves of these plants can develop in different ways, depending on whether they are surrounded by water or air. Leaves produced lower down on the stem, in the aquatic environment, tend to be finely

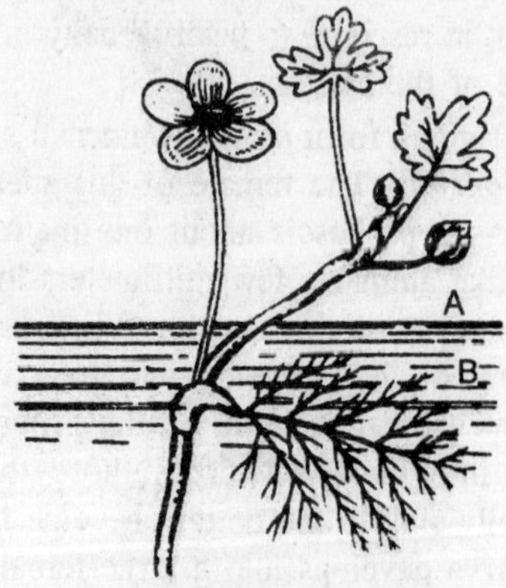

Fig. 7.1. Ranunculus aquatilis water foot. A—The part developing lobed leaves; B—The part developing filamentous prickles.

dissected and feather-like, whereas those borne in the air are more compact. The plant effectively displays the height of the water in its own structure. This reflects the way the developing leaves respond to their environment. If a young plant growing completely under water is moved to a dry environment, it will switch from producing feathery leaves to aerial leaves. Conversely, a plant that has been raised entirely in the air and then submerged in water will start to produce leaves with a feathery shape. In each case, the previously fully grown leaves do not change; it is the newly arising leaf-buds at the tip of the plant that develop according to their new surroundings. This response allows the shape of the developing leaves to be adjusted according to their environment: large feathery leaves are appropriate for waving about in water whereas compact leaves are more suited to aerial surroundings.

The development of organisms may also respond in various ways to physical damage inflicted by their environment. There are two broad types of responses. One of these as *restoration*, and the other as *regeneration* (sometimes both of these processes are put together under the general heading of regeneration).

Restoration

The propensity for *restoration* following damage is particularly well developed in plants. When you mow the lawn or prune a bush, the plants usually manage to recover.

Unlike most animals, which have a more defined endpoint in development, plants continue to grow and add new parts to themselves throughout their lives. This is achieved by groups of cells, called meristems, that continually divide and replenish themselves whilst at the same time adding more tissue to the plant. *Meristems* can be thought of as being perpetually embryonic cells, which, like Peter Pan, can continually maintain an early condition rather than maturing. For example, there is a primary meristem at the main growing tip of a shoot, which replenishes itself whilst also adding cells to the stem below and to leaf-buds on the side. The primary meristem continues to generate leaf-buds as it climbs upwards, so you end up with a series of leaves of different ages, with the youngest at the top and progressively more mature stages as you go down the plant. The *primary meristem* also produces additional meristems, called *secondary meristems*, on its periphery. Typically, a secondary meristem is to be found in the angle between each leaf-bud and the stem. Each secondary meristem can be thought of as a very young shoot-bud. Once it starts to grow, the secondary meristem will itself produce stem, leaf-buds

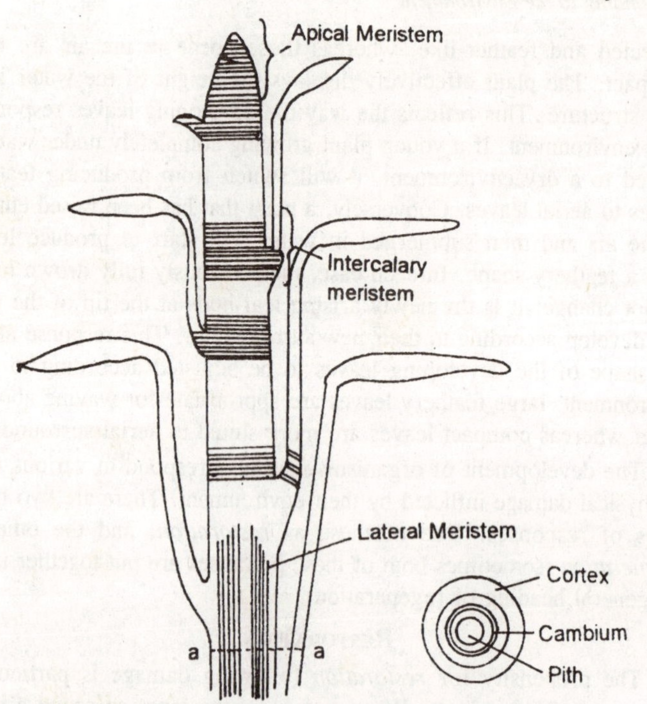

Fig. 7.2. A—Diagrammatic L.S. of shoot apex showing the position of meristems; B—T.S. of shoot at A-A.

and yet more meristems (flower-buds may also arise from meristems but their growth is more limited). Potentially the process could carry on indefinitely, with the plant becoming more and more branched. In practice, however, many of the additional meristems and buds tend to remain dormant for long periods, so that growth is channelled in a limited number of directions.

When plants are cut or pruned, they typically respond by activating growth from meristems and buds that were previously dormant or slow-growing. The system ensures that when a major growing point is lost, others are activated and take over. Similarly, when you mow the lawn, it does not grow back from the cut surfaces of the leaves, but from meristems and buds that were already present in the grass plants, below the level of the cutting blade. That is why you should not set your blades too close to the ground, as you will then remove the buds and meristems. These are all examples of response to environmental damage through the development of meristems and buds that were already present.

There is a somewhat analogous situation in some animals. In the 1740s, a small freshwater animal was the talk of the scientific world. Abraham Trembley, working in Holland, had come across a small animal that could keep producing entirely new individuals after it was cut up. The animal had a relatively simple organization: it comprised a trunk with arms or tentacles at one end (head) waving about in the water, and a disc at the other end that attached it to the substrate. When Trembley cut the animal in half to see whether the separate pieces could survive on their own, he was astounded to see that both halves started to develop into new animals. He had managed to produce two complete animals by cutting one in half! Eventually, the animal was named *Hydra* after the monster in Greek mythology that could produce new heads from the trunk every time that Hercules cut one off. Trembley's *Hydra* was even more remarkable than the monster because each removed head could also produce an entire animal.

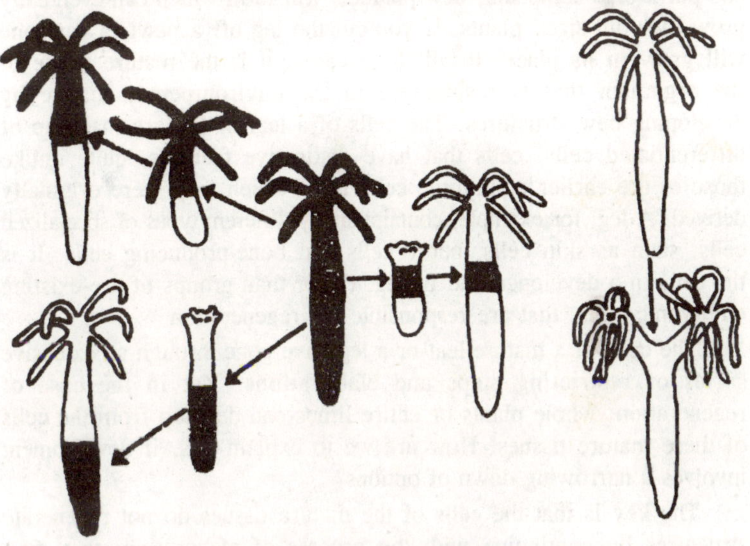

Fig. 7.3. Regeneration of Hydra.

The ability of *Hydra* to behave like this reflects the way it normally grows and propagates itself. A well-fed Hydra will form a bud about two-thirds of the way down the trunk. The bud then elongates and sprouts arms at the end. In two or three days, the bud looks like a little *Hydra* and can then pinch off at its base to become an independent individual. Like plants, *Hydra* can grow from buds; but whereas plants produce their buds from meristems at the growing tips, in the case of

Hydra the buds are produced directly from the trunk. The trunk can be considered as equivalent to a plant meristem, the cells maintaining themselves in a somewhat embryonic condition while at the same time producing buds to the side, arms and mouth at the head end, and a basal disc at the other end. Seen in this light, slicing up *Hydra* is equivalent to pruning a plant. Cutting the animal stimulates the production of new structures from the embryonic cells in the trunk, much as pruning a plant can stimulate the development of pre-existing meristems.

Regeneration

Another type of response to damage is *regeneration*. This involves the production of new structures from mature tissue, unlike restoration, which comes from pre-existing cells of a more embryonic nature. For example, if you take a leaf from an African violet or begonia plant and put it onto some soil, new plantlets will form which can eventually grow into full-sized plants. If you cut the leg off a newt, a new one will grow in its place. In all these cases, it is the mature tissue of the organism that is responding to the environmental change by developing new structures. The cells of a leaf or leg are made up of differentiated cells: cells that have distinctive features, quite unlike those of the earlier embryonic cells from which they were originally derived. A leg, for example, contains many different types of specialized cells, such as skin cells, nerve cells and bone-producing cells. It is the cells in a developed leaf or leg, rather than groups of pre-existing embryonic cells, that are responsible for regeneration.

The cells of a mature leaf or a leg have gone through an extensive series of restricting steps and elaborations. Yet in the case of regeneration, whole plants or entire limbs can develop from the cells of these mature tissues. How are we to explain this, if development involves a narrowing down of options?

The key is that the cells of the mature tissues do not regenerate structures by continuing with the process of elaboration; they first have to go back to a more embryonic condition. Because patterns are refined through building on previous steps, the only way to get regeneration from cells in mature tissue is for them to return to an earlier state of development, and start to rebuild the picture all over again. This is precisely what seems to occur. When a begonia leaf is cut off a plant and placed on some soil, its cells do not directly switch to become root or shoot cells. They first become dedifferentiated, going back to the cell types typical of embryos or meristems. Only after this do they start to grow and divide again to form a new plant. They have to go back before they go forward again, as might be expected from a system which builds on a sequence of steps.

Some plants exploit this as a normal way of propagating themselves. In mature leaves of the bog orchid Malaxis paludosa, the cells at the tip dedifferentiate—they return to a more embryonic condition—allowing them to develop into tiny egg-shaped embryos. When these embryos become detached from the leaf, they develop into new plants, just like the embryos produced through seed (although unlike plants derived from seed, the new plants are genetically identical to the parent plant because no sex is involved). In all these cases of regeneration from mature tissues, the cells must first return to a more embryonic condition before they can start building up their internal patterns again.

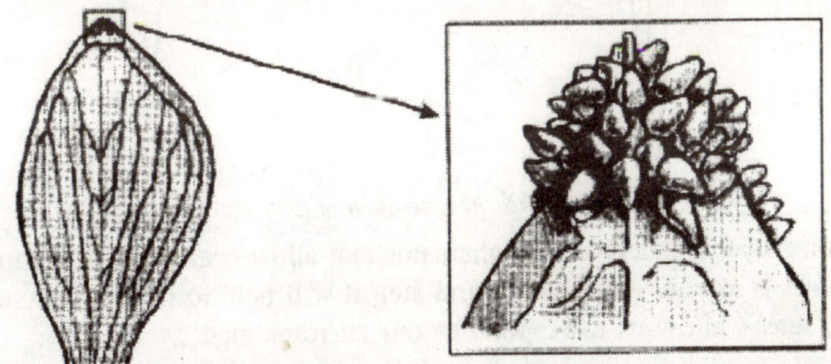

Fig. 7.4 Leaf of Malaxis paludosa showing the development of a cluster of embryos at the extreme tip.

The same applies to regeneration in animals. When a newts limb is severed, the stump heals over and cells near the tip start to lose their characteristic features. Bone cells, cartilage cells, nerve cells and so on lose their particularities and become a mass of dedifferentiated cells. These cells then start to grow and develop into the missing part of the limb. The new limb does not come from cells of the stump directly changing to those of a new limb. The cells go back to an earlier, more embryonic state, and the limb then regrows just as it did when it first developed. As with regeneration in plants, the cells have to go back before they can go forward.

A Sense of the Surroundings

These examples have been quoted to show how the development of organisms can respond in a variety of ways to external influences such as temperature, light, gravity, water and physical damage. Look

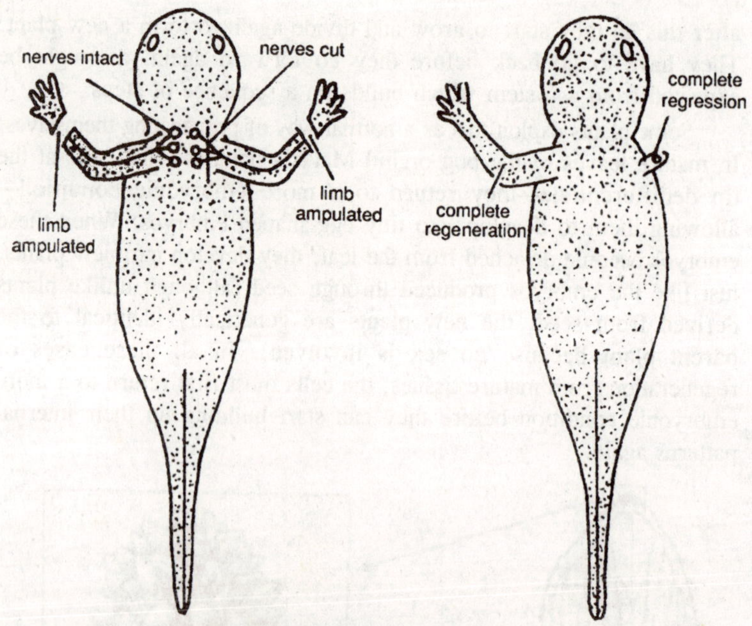

Fig. 7.5. Regeneration of limb.

at some of the underlying mechanisms that allow organisms to respond to what is around them. As a first step it will help to look at how our own senses allow us to respond to our surroundings.

We saw in the previous chapter how our senses of smell and taste depend on receptor proteins that are triggered by molecules in the environment. The receptors allow our internal chemistry, the reactions within our cells, to respond to chemical information from outside. Our other senses, those of vision, hearing and touch, also depend on receptor proteins, but instead of responding to molecules in the environment, they are triggered by other factors. The receptors convert stimuli such as light, sound or pressure into the chemical language of cells. The net result is the same as for smell and taste—the internal reactions of cells are altered—but the stimulus is different.

Our sense of vision provides a very good example of how this works. There are a particular group of receptor proteins, called *opsins*, in the cells of the retina at the back of the eye. Each of these receptor proteins changes shape when it is struck by a particle or *photon* of light (to be precise, it is a small molecule attached to the receptor protein that changes shape when it absorbs the light, and this then leads to an alteration in the shape of the receptor protein). The altered shape of the receptor protein then sets off a chain of reactions in the cell. The receptor protein effectively converts a physical signal from the outside, *light*, into the language of chemical reactions within a cell. These chemical reactions in the cell in turn lead to charged *atoms* (ions) moving across the cell membrane, resulting in an electrical signal, a *nerve impulse*, being sent to the brain. We can summarize by saying that a light stimulus is converted into a chemical reaction which then leads to an electrical impulse.

We have four types of light-sensitive cell in the retina of the eye, each containing a characteristic type of receptor protein. Cells of one type, called *rods*, contain one sort of receptor protein and can be triggered by low levels of light. These are the cells that allow us to see in dim light. The three other types of cell, called *cones*, are responsible for colour vision in bright light. Each of the cone cells contains receptor proteins that respond to one of three colours: *blue*, *green* or *red*. The combination of cone cells stimulated in an area Of the retina eventually leads to what we perceive as colour. Our ability to see anything therefore depends on these various types of receptor protein. Without them, we would be completely blind. Colour-blindness results when one of the genes for the receptor proteins is missing or defective (red-green colour-blindness is particularly common in males because the genes for the red and green receptor proteins are on the X chromosome. Males have only a single X chromosome in each cell, so if this chromosome carries a mutation in one of these genes, the individual will be colour-blind. Females have two X chromosomes, so even if one of these carries a defective gene, the other X chromosome will most likely carry a normal copy that acts as a backup.)

Our sense of touch and hearing are less well understood than vision, but they also depend on receptor proteins. In the case of hearing, receptor proteins in the sensory hair cells of the inner ear are stimulated by sound vibrations. This triggers a change in their shape, leading to nerve impulses being sent to the brain. Similarly, our sense of touch and muscle movement depend on receptor proteins that alter their shape in response to mechanical changes, such as stretching of the cell membrane.

We sense our environment through receptor proteins, proteins that change their shape in response to chemicals, light, sound or other stimuli. They convert information from the outside into the internal chemical language of cells, eventually leading to nerve impulses being sent to the brain.

Modifying Internal Patterns

In many ways, the principles by which developing organisms sense their environmental conditions are not too different from the way our own senses operate. In both cases, receptor proteins of various types play a key role in responding to external circumstances. The main difference has to do with the outcome of triggering the receptors. In the case of our senses, a nerve impulse is eventually sent to the brain, whereas for development we shall see that the outcome is often a modification of the organisms hidden colours.

We have already encountered a situation in which the triggering of receptors can lead to a change in internal patterns. Recall that in the previous chapter, we saw that cells can respond to scents from their neighbours by changing their hidden colours. The scents trigger the receptor proteins, setting off a chain of reactions in the cell which eventually lead to an alteration in the shape of a master protein, a change in hidden colour. This new colour can then be interpreted by other genes in the cell which may get switched on or off as a consequence. In a similar way, the receptor proteins triggered by the external stimuli, such as light, water or physical damage, may eventually lead to a modification of the organisms hidden colours. This change can then be interpreted by various genes, leading to a change in the proteins being produced in the cells and hence their properties. There are still major gaps in our knowledge about how this occurs, but it is worth looking at a few examples to illustrate the likely principles involved.

How do plants respond to light? Plant cells are known to contain several different types of receptor protein that are sensitive to light. Like the receptors in the eye, these plant receptor proteins change shape when light strikes (as with the eye receptors, they have a small molecule attached to them that changes shape when it absorbs light; this influences the shape of the receptor protein). This change in shape of the receptor protein then sets off a chain of events in the cell. We still do not know the details of all these events, but one important outcome appears to be a change in the cell's hidden colour. *Genes* may interpret the altered colour in various ways, being expressed or not as the case may be, eventually leading to a change in cell properties.

It shows what typically happens when a plant tip that was originally growing upwards is illuminated from only one side. The cells on the lit side grow more slowly than those on the shady side, resulting in

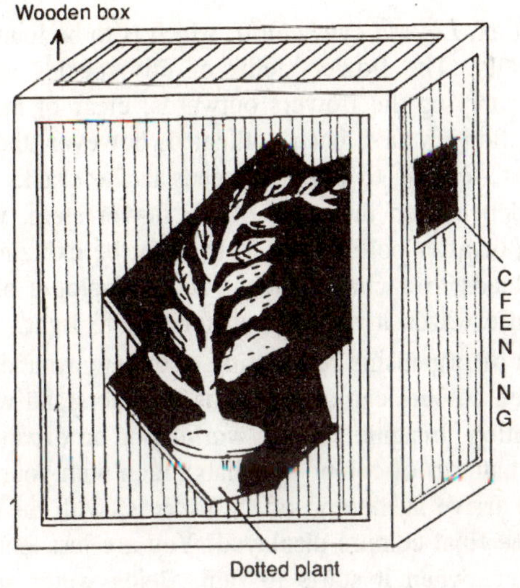

Fig. 7.6. Phototropic chamber.

the stem bending towards the light (in some cases, cells on the shady side may also grow more quickly). All the details of how this response occurs are not known, but let me give a simplified explanation of what may be going on. Because the plant is illuminated from only one side, the light receptor proteins on the lit side of the growing tip will be triggered more than those on the shady side. This sets off a chain of chemical events which may eventually result in a difference of hidden colour between the two sides, with the cells on the lit side turning from, say, cloudy-grey to sunny-yellow. Various genes then respond or interpret the sunny-yellow colour in such a way that the proteins they produce result in growth slowing down. This means that the cells on the lit side grow more slowly, causing the shoot to bend towards the light.

The way the system behaves depends on the particular combination of hidden colours in the responding tissue. For instance, in the previous example, the shoot started with one overall colour, cloudy-grey, which responded to the light stimulus by turning sunny-yellow. If the tissue had started with a different set of colours, as for example would be the case in a root tip, it may have responded entirely differently, say bending away from the light rather than towards it. This is because both the sensitivity (receptors) and response will depend on what the hidden colours are to begin with. This allows responses to change at different times, according to the hidden colours. A good example is in ivy-leaved toadflax, *Linaria cymbalaria*, which is to be found commonly growing on walls. The flower stems of this species initially grow towards light, carrying the flowers outwards, clear of the leaves and wall. After the flowers have been pollinated, however, the stems twist away from light, tending to bury the seeds in the cracks of the wall. The response depends on the stage of the flower stem, whether it is reacting before or after pollination. This might be explained if the act of pollination leads to a change in the hidden colours of the flower stem, so that afterwards it responds differently to light.

The notion of organisms responding to the environment through a change in hidden colours can also help us to understand what happens during regeneration. Imagine you are working on an expanding canvas, elaborating the patterns of colour in collaboration with your neighbours. Eventually you arrive at the end of the process, with the canvas fully extended and the final colours displayed. You are just feeling satisfied that it's all over, when it starts to rain. Being water soluble, your colours are washed away and you react by starting the whole painting exercise again. In a similar way, when the cells of the begonia leaf are presented with a new environmental situation, as when they are cut off a plant and plunged into soil, their hidden colours return to an earlier state. It is not that the environment actually washes away the hidden colours, but particular cells in the leaf respond to the new situation by modifying their colours so that they eventually resemble those of an earlier stage.

Limb regeneration can be explained along similar lines, with the cells in the stump responding to the operation by returning to producing an earlier set of hidden colours. In this case, however, there is a further feature that needs to be explained. Only the part of the limb that is removed tends to be regenerated. If a newt's limb is amputated near the base, almost the entire limb is replaced, but if the amputation is at the wrist, only the hand regenerates. As noted by the eighteenth century French scientist Rene-Antoine Ferchault de Reaumur, who was studying the same phenomenon in crustaceans: 'Nature gives back to the animal precisely and only that which it has lost' How is it that the dividing cells only regenerate what is missing and not more or less? The answer to this is not known, but one possibility is that it has to do with the persistence of some hidden colours in the limb. We might imagine the newt limb is divided into a series of hidden colours from base to tip during early development, and that these colours are then maintained even up to the time that the limb matures. When the limb is amputated, the cells at the tip of the stump may retain some of these hidden colours, even though others are lost or modified. This means the regenerating cells in the stump will have a different set of hidden colours when the amputation is at the wrist, compared to when it is carried out near the base. This could provide some of the information that determines how much of the limb is then regenerated.

In all these cases of regeneration, the cells respond to an injury by changing their hidden colours to resemble an earlier state; but there is also an artificial method of achieving a similar sort of result. This involves taking a nucleus from a cell in the mature body and transplanting it into an egg cell that has had its own nucleus removed; equivalent to physically transporting an artist from a mature canvas onto a fresh piece of initial canvas with the early hidden colours. Being surrounded by this new cellular environment, the nucleus might then start to interact with it in such a way that a new individual eventually develops. The new individual would carry exactly the same set of nuclear genes as the individual that donated the nucleus: it would be a clone of the original (just as a plant derived from regeneration of a begonia leaf is a clone). This is essentially the way that researchers were able to clone a sheep, called *Dolly*. A cell from the mammary gland of a mature sheep was fused with an unfertilized egg cell that had had its own nucleus removed. In this case, the cytoplasm from the donor cell was introduced together with the nucleus, but the effects of the donor cytoplasm were presumably swamped out by the cytoplasm in the large recipient egg cell.

Hormones

An important feature of many of these environmental responses is that they are often coordinated through long-range signalling between cells. This can be achieved through a set of internal scents, called *hormones*. Hormones are typically small molecules (small, that is, relative to most proteins), many of which are synthesized through a series of chemical reactions catalyzed by proteins (*enzymes*). As with other types of internal scent, these hormones match the shape of particular types of receptor protein, allowing them to trigger reactions in cells which may lead to a change in hidden colour. However, unlike most of the other scents, they can sometimes travel relatively large distances within the organism.

Plant hormones, sometimes called *growth regulators*, have been shown to be involved in a range of environmental responses including stem growth, flowering time, fruit ripening, leaf shape and leaf fall.

In many of these cases, their role seems to be to coordinate the responses between cells in the plant. A good example is what happens in pruning. If you remove one or more growing tips of a plant, it will stimulate the shoot-buds further down to develop and grow out. An injury caused by the environment in one part of the plant influences growth and development of structures that are very far away in terms of cell numbers. This effect is mediated by a particular growth hormone, called auxin, that is produced at the growing tips. *Auxin* travels from the tips down the plant, where it inhibits the growth of side buds. That is to say, the cells in the side buds respond to the auxin scent by slowing down their growth. Cutting off a tip removes a major source of *auxin*, liberating the side buds from its inhibitory effects, so they start to grow out (a similar sort of explanation accounts for the response of Hydra to surgery).

Flowering time provides another instance of long-range communication. Many plant species flower under particular day-length conditions, ensuring that they produce flowers in one season. Some species (e.g. *chrysanthemums*) flower when days are short, in autumn, whereas others (e.g. *antirrhinums*) flower when days are long, in summer. The response to day-length depends on receptor proteins in the leaves that are triggered by light. When the duration of the day reaches a certain length, it sets off a chain of reactions in the leaf cells. As a consequence, one or more long-range signals are then produced which travel up to the growing tip, triggering cells there to initiate flower development. The precise nature of these signals has yet to be determined, but some hormones have been strongly implicated.

Hormones also play a role in coordinating the way animal development responds to its environment. For example, the effect of temperature on the determination of sex in reptiles is coordinated by hormones in the developing animal. Each sex produces different hormones. The cells of the animal then respond to the hormones, leading to male or female characteristics. The precise mechanism by which temperature influences sex hormones in reptiles is not known, but one model is that it changes the hidden colours of the cells in the developing gonads, which then leads to altered levels of hormone synthesis. (The sex of mammals is also coordinated by hormones, but in this case the type of hormone produced depends on the presence of a gene on the Y chromosome rather than temperature. This gene codes for a hidden colour and is expressed in the developing gonads of the male, where it leads to testis development and the production of the male sex hormone, testosterone. Females do not have a Y chromosome, so their gonads do not express this hidden colour and therefore develop into ovaries, which do not make testosterone.)

Another example of how hormones can coordinate responses involves moulting in insects. The growth of insects is largely limited by their rigid outer covering, which must be shed every so often if they are to continue growing. In the blood-sucking bug *Rhodnius*, this moult occurs after the animal has gorged itself on a large blood meal which distends the abdomen. The distension stimulates receptors in the cells of the abdomen which then send nerve impulses to the brain. This in turn leads to the insect producing a small molecule, a hormone called ecdysone, from one of its glands, which then circulates throughout the body. Cells respond to this internal scent by changing their hidden colour, switching on the genes needed for moulting. The role of the hormone is to coordinate changes in hidden colour throughout the body in response to the consumption of a large meal.

Responding in Art and Development

These various examples of environmental responses show that the elaboration of hidden colours, scents and sensitivities does not occur in isolation but is influenced by the general surroundings. In all of these cases, two sorts of process need to be borne in mind.

On the one hand there is the internal process of elaborating hidden colours. This involves a highly interactive series of events in which local signalling between cells, through scents and sensitivities, allows patterns of hidden colour to be built upon and refined in a coordinated way. In this case all the signals are generated by the organism itself. This is comparable to a painter interacting with a canvas, where all the colours are applied by the artist. The artist is responding to signals that he or she has put there.

On the other hand, this system of internal elaboration is continually responding to the external environment. In the case of development the mechanisms are essentially the same as those for internal patterning, the main difference being that the receptor proteins are triggered by signals from the external environment, such as light or gravity, instead of scents produced by other cells. Unlike the process of internal elaboration, the signals in this case are not generated by the organism itself but are relatively independent of it, coming from its surroundings. In a similar way, the dialogue between the artist and canvas is influenced by the general surroundings, such as the lighting conditions or landscape. The artist detects these environmental factors in the same way as the colours on the canvas, through the visual senses. The difference is that in the case of the canvas, the stimuli are a direct consequence of the artist's actions, the colours applied to the support, whereas the external environment is more independent of the artist.

For many aspects of development, particularly in animals, the environment may be held relatively constant, providing a uniform context that allows development to proceed. *Heterogeneity* in the surroundings plays a relatively minor role in the patterning process, as is the case with an abstract painting.

By contrast, plants are the landscape artists of the living world. Plants are continually modifying their hidden colours in response to variation in their surroundings. Like artists, they alter their patterns in response to the light and shade around them. They also establish orientations of colour in response to gravity, much as an artist might emphasize vertical or horizontal lines in a painting. Amphibious plants even modify their hidden colours according to the level of water in a pond, just as someone might depict the surface of water in a lake-side scene. The environment is written all over a plant, not as a mould over a cast, but through the particular ways that the plant responds to its circumstances. Some of the more obvious examples to make this point clearer but there are many other more subtle ways in which plant development is continually reacting to its surroundings. If we could read plants as well as we read pictures, perhaps we could tell as much about an external environment from a plant as from any landscape painting.

Developing Brain

Development in plants responds more to variation in the environment than is the case in animals. Yet it seems to me that there is a very important aspect of animal development that comes much nearer to the plant style: brain development. On the face of it, it seems peculiar to make such a comparison, because brains are what we least associate with plant life. But if we look at the problem in terms of how processes are influenced by the environment, it is in brain development that responses to the surroundings start to become very significant for animals. Although we are still very far from understanding how the brain develops, it is worth pursuing this comparison further so as to bring together some of the key issues raised so far in this book.

Our brains contain many billions of nerve cells or neurons, which are connected with each other to form complex networks of interacting cells. The neurons work by transmitting electrical impulses along their length, which then influence other neurons that are connected to them. These connections are not direct: there is a tiny gap between each neuron and the next, called a *synapse*. When an electrical impulse from one neuron arrives at a *synapse*, it causes the release of a specific scent (signalling molecule), called a *transmitter*, which then moves across the gap, triggering a receptor protein in the neuron on the other side. This in turn sets off an electrical impulse in the responding neuron, allowing further transmission of the signal. It is the pattern and strength of all the connections between the various neurons in the brain that is believed to underlie our thought processes.

Many of the basic features of our brain's layout are established during development in the womb, through a process of internal elaboration that probably has much in common with many of the other developmental systems. However, the patterning process does not stop there. As soon as we are born, the brain starts to get bombarded with information through our various senses. As a result, the pattern and strength of connections between nerve cells in the brain become modified in response to experiences. We normally call this *learning*, rather than development, because the environment plays such a large part. But we might equally say that it is simply a more plant-like form of development, in which the organism modifies its internal patterns in response to variation in its surroundings. The internal patterns here refer to the set of connections between the cells of the brain.

Let me give an example to illustrate the point. If one eye is kept covered at a critical period within the first few months of birth, so as to deprive it of any visual stimulation, the neurons emanating from that eye will fail to form connections with the main brain— so that the eye becomes almost entirely and irreversibly blind, while the other eye is still perfectly fine. In other words, without stimulation from the outside, the connections from the eye to the brain do not form. This mechanism is thought to ensure that only active or working neurons that come from the eyes are connected to the brain. By covering one eye, the neurons that come from it essentially behave as defective neurons, and so are not connected up.

This case shows, in a somewhat extreme way, how the patterns of cell connections that form in the brain may be influenced by the environmental conditions. Because the resulting change is so severe in this case, involving the loss of sight from one eye, we would probably say that the environment is affecting the development of the brain.

But we might also say that the young mammal is learning about its environment, forming the appropriate connections in the brain that allow each eye to function properly. In more subtle types of response, as when the cell interactions in the brain are modified by being exposed to language or objects around us, we would call the process learning. But we could equally say that it is a form of development in which the internal patterns of cell connections are being modified through a response to the surroundings. The point is that our brain processes are not divorced from development, they are a rather special extension of it.

As with other examples of development, we can distinguish between various styles in which the brain may interact with the environment. In a uniform environment, as when we are in a quiet dark room, the brain is still very active, with one set of cell interactions leading to another; this is comparable to a developing organism elaborating its hidden colours in a relatively constant and uniform environment. We could call this pure thinking rather than learning, because of the lack of stimulating input from the outside.

When we are exposed to a stimulating environment, we talk about learning from our surroundings rather than just thinking. It is not that thinking stops but that it now also responds to numerous environmental stimuli. As with other cases of development, learning is not a matter of the environment moulding or imposing itself on the individual. Our brains are not like buckets that are passively filled up with environmental information; rather they respond in particular ways to their surroundings. Stimuli from the outside trigger receptors in our various sense organs, leading to a complex response. The response will depend on the particular history of the brain, the way it has developed and interacted with the environment on previous occasions. This is comparable to plant development, where internal patterns change in response to variation in the environment.

In a sense, both thinking and learning are creative activities: they are highly interactive processes that lead to novel ideas or levels of understanding by building on what went before. These activities are based on the history of each individual, the way his or her brain has developed and responded to previous experiences. This is true whether we are talking about a child learning how to read, or a poet making up a new verse. In all of these cases, the outcome is not anticipated in advance: the child does not know what it is like to be able to read before having learned, and the poet does not know the new lines before he or she has composed them.

The same is also true for the way that theories arise in science. As emphasized by the philosopher Karl Popper, scientific hypotheses or theories are not arrived at by following a plan or logical procedure, they arise creatively: '... theories are seen to be the free creations of our own minds, the result of an almost poetic intuition, of an attempt to understand intuitively the laws of nature.'

A scientist does not derive theories by following a formula, they are the product of a freely creative mind. By free, Popper does not mean to imply that the theories come from nowhere: theories depend on the scientist's pas t, including earlier observations and theories, going all the way back to childhood. When a scientist sitting in a bath suddenly has a new idea, it may feel as if it has come out of the blue, but it is still based on the previous history of the individual scientist. Popper's point is that a scientist does not come up with a new theory by a logical analysis of this earlier information; rather, the theory arises through a creative process in which his or her brain builds on what went before. It is only after the scientist has come up with a hypothesis, that logic comes into play, through testing its predictions against observations. This may lead to the rejection of the hypothesis because it is found to be logically inconsistent with the observations, or perhaps the hypothesis survives the test and is therefore corroborated. In either case, the hypothesis itself is not derived by a logical procedure but by a creative process in the brain.

Now there is a further aspect of environmental interaction that starts to become increasingly important in the case of the brain. The general environment as something that is relatively independent of the developing system. In most cases this is only an approximation because the individual does influence its surroundings to some extent as it develops. In the case of the brain, this can be quite significant. In addition to receiving various stimuli, it also influences the external world through the way we act. When we see an apple on a table we may respond by picking it up and eating it. Instead of the apple being stationary, we see it moving towards our mouth and start to detect its smell and taste. This modification in the environment, the movement of the apple, is a consequence of the brains original response, the desire to pick up the apple in the first place.

If the extent of these two-way interactions between the brain and a part of its environment become particularly pronounced, then this aspect of the environment can no longer be considered as truly *external* or *independent*; it becomes integral to the interactive process. This is precisely what happens in the case of painting. The artist and the canvas are so interdependent that the canvas becomes almost an extension of the artist. Rather than the canvas being an *external environment*, it becomes an *internal environment* that is continually being modified as well as responded to. Although the canvas is physically outside the artist, it has been internalized within the creative process. It is as if the brain has established a two-way interaction to such an extent that its dynamic processes spill onto the canvas.

To my mind, human creative processes, such as painting, thinking and learning, have much in common with development, in the sense that they all depend on informed responses that continually elaborate what went before. In each case, we do not need to invoke instructions that are being followed, software as distinct from hardware, or plans that are subsequently executed. A further aspect of all these processes is that each responds in some way to a general environment. In some cases, as with pure thinking, abstract painting, and for many aspects of development, the general environment provides a relatively uniform background that allows the process to occur. In other cases, as with learning, some aspects of plant development, and landscape or portrait painting, heterogeneity in the general environment plays a more important role. Here, the creative process is continually responding to variation in its surroundings. In none of these cases does the environment mould or dictate the way things should go; rather, the creative process is modified to a greater or lesser extent in response to what is around it.

The reason for this commonality between human creativity and development is that the brain is itself a special extension of development. The interactive processes that characterize development themselves underlie the way our own brains work.

This comparison works so well is that creativity is itself grounded in developmental processes: creativity is the child of development. This is not to deny the importance of the environment, because as we have seen, the environment plays a fundamental role in development. The brain is the same as any other group of cells; it clearly functions in a distinctive way through the complex interactions between neurons. My point is that there is a continuity between the process of development and the functioning of the brain, which leads to them having many features in common: both are highly interactive processes that elaborate on what went before in a historically informed manner, in the context of a particular environment.

Once this is appreciated, it becomes possible to look at our own creativity in a new light. The interactive processes within the human brain continually manifest themselves to us in terms of our creative thoughts and activities. But we cannot access the processes underlying these experiences purely by introspection. We cannot unravel the mechanisms behind our thoughts simply by thinking about the problem, because each thought we have will itself be a manifestation of the underlying process. This scientific understanding is itself dependent on our own creativity: our ability to come up with ideas and hypotheses. But these hypotheses have the merit of being testable through experiment and observation, allowing us to get a distinctive insight into the problem. Our scientific understanding of development can give us some useful intuitions about ourselves: we can begin to see how our own creativity might arise as an extension of an underlying developmental process, rather than as something that floats all by itself.

8

EVOLUTIONARY FACTS

Imagine yourself living in the two decades after the publication of the *Origin*. You have read Darwin's book as well as reviews, heard several lectures on the subject, and possibly even attended the Oxford debate between Wilberforce and Huxley. If you were not a scientist and were unfamiliar with the workings of scientific procedures, you might be somewhat puzzled and feel unsatisfied with the data available to Darwin and other naturalists that were regarded as "*proofs*" of evolution. You might have thought that evolution meant the conversion of one species into another, as the title of Darwin's book suggested; yet not a single example of this conversion was provided in the *Origin*. Instead, the proofs of evolution consisted of a heterogeneous collection of data: the finding that fossils and living species of armadillos of Argentina are very similar; the restriction of related species in the Galapagos to their own islands; the recapitulation of reptilian jaws in mammalian development; the different kidneys in embryos and adults of vertebrates; vestigial organs such as the human appendix. Would you have found the Origin convincing?

The proofs of evolution do not come from the experimental demonstration of one species changing into another. Such direct evidence will always be absent except for unusual situations because the change of one species into another may take many thousands of years—and none of us is able to wait that long for definitive proof. Instead, we have to put together bits of evidence that, with luck, will provide a plausible explanation of what occurred. A detective may study the grooves on the murder bullet, trace the activities of suspects, wonder why the household dog did not bark, compare the DNA in a drop of blood at the murder scene with that of suspects, and seek a motive.

Neither the detective not the evolutionist observes the events that require explanation. Both, however, proceed in a similar manner and reach a highly likely conclusion.

The data that prompt biologists and paleontologists to accept evolution as the most accurate statement possible about the diversity of life are indirect, but each piece is consistent with the concept of evolution. Although we cannot "*see*" evolution on any major scale, much of the data of biology and paleontology cannot be understood without it. Furthermore, the theory of evolution continues to suggest new ways of obtaining further understanding of the origin and diversification of life over the ages. There is no other scientific theory that has proven to be as useful, and for this reason evolution is now accepted as true beyond all reasonable doubt.

Evidence for most of the data that convince scientists is rarely convincing, or even understood, by nonscientists. For example, most people accept what astronomers tell us—that the rotation of the Earth on its axis is the cause of night and day—but can laypeople recite the evidence that this is so? Not likely. To take a more recent example, a layperson who is shown the data on the temperatures in deep space and the change in the spectral lines of light from distant galaxies probably cannot figure out that these data are evidence for the Big Bang that started the universe 12–15 billion years ago. If a nonspecialist accepts the *Big Bang* as a useful concept, it is because he or she has confidence that astrophysicists are giving the best explanation they can on the basis of available evidence.

There is no strict way scientists—or detectives—go about their quests for answers. There is no one rigid scientific method, but scientists do follow in a general way a series of steps. First, they formulate a question about some natural phenomenon that needs to be explained. The one evolutionists start with is: How can we provide a scientific explanation for the many kinds of organisms that lived in the past and are alive today? The next step is to guess what an answer might be. That guess, or hypothesis, is almost always based on preliminary observations from nature or data from experiments that make the hypothesis plausible. In Darwin's case, he based his hypothesis on his observations of geographic variation among animals and plants in the Galapagos and on the resemblance of fossils in Argentina to living species found there. Both observations could be explained by the hypothesis that evolution works through natural selection acting on the variation present in populations.

The next step is to test the hypothesis to see if it can explain other kinds of data. This is done by deducing the consequences that would occur if the hypothesis were correct. A deduction, then, is a logically necessary derivative from the hypothesis being tested. If the hypothesis is that the reptiles evolved from the amphibians, an important deduction would be that the fossil record should contain individuals that are intermediate between the amphibians and their descendants, the reptiles. Since both are bony creatures, and since bones have a good chance of being fossilized, with enough luck and labour fossils of organisms intermediate between amphibians and reptiles should be discoverable.

If a fossil structurally intermediate between an amphibian and reptile were discovered, the cry "*Eureka*!" would be premature unless another deduction—that the intermediate organism must have lived after the amphibians first appeared—were also found to be true, since descendants cannot be more ancient than their ancestors. To test this deduction calls for reliable methods for determining the age of sedimentary rocks in which fossils are entombed. Just knowing the relative ages of the strata is often satisfactory for this purpose. For example, if all of the intermediates between amphibians and reptiles were in strata younger than those containing the earliest reptiles but older than those with the earliest amphibians, the hypothesis would be substantiated.

A fruitful scientific theory is one that

1. explains a natural phenomenon with logical and internally consistent arguments
2. relates that explanation to the existing conceptual scheme
3. bases that explanation on confirmatory data from observations and experiments
4. rigorously excludes supernatural phenomena as explanations
5. reduces the complexity of nature to relative simplicity
6. is intellectually satisfying and even elegant
7. suggests experiments and observations that expand the implications of the theory and increase the probability that it is not incorrect

Instead of "increase the probability that it is not incorrect," why not "increase the probability that it is correct"? The reason is that a single experiment or observation invalidating a scientific statement means that the statement must be modified or abandoned. Suppose the hypothesis to be tested is "*Alice* is six feet four inches tall and is thus the tallest woman in the world." One could test this hypothesis by comparing Alice with dozens of other women and probably find that she is taller than any. So far so good. But to prove it correct, one would have to measure all of the women in the world. In an actual experiment, it would probably be necessary to check the heights of only several thousand other women to find one taller than six feet four inches. Thus, a single observation would disprove the hypothesis, but an impractical number of observations—checking all women in the world—would be required to prove it correct. A hypothesis in science remains useful if an ever-increasing number of observations seem to indicate that it is correct and no observations prove that it is wrong. Nevertheless, the most accurate comment about any scientific statement is that it not be accepted as true in any final sense but that it be tested repeatedly, never falsified, and so remain for the moment true beyond all reasonable doubt.

This stricture, to never say that a statement in science is true, is philosophically correct, but in the real world such skepticism can border on nonsense. Science does work: rockets do reach the moon and planets; the positions of the major objects in the solar system can be predicted with great accuracy; human diseases can be prevented, cured, or lessened in their severity; and a steady stream of technological wonders emerge from our laboratories and factories, based on knowledge that practically speaking can be accepted as "*true.*" Scientific knowledge may not be true in a philosophical sense, but it is often quite adequate to provide us with an extraordinary level of understanding of ourselves and our world.

Search for Missing Links

Darwin's theory of evolution implies that although at the microevolutionary level the changes in a lineage are gradual, the summing of these small steps over time results in differences so great that descendants and their ancestors fall into different taxonomic groups: at first, different species, then genera but, with more time and evolution, different families, orders, classes, phyla, and kingdoms.

If this is indeed the way evolution works, surely some fossils intermediate in structure between two major groups, such as fishes and amphibians or reptiles and birds, could be found. In Darwin's day these intermediates were called *missing links* because for so long they were indeed missing. Darwin could not provide a single example of this kind of intermediate, such as between two classes of vertebrates. Needless to say, the antievolutionists argued that the absence of this critical information made the hypothesis of evolution improbable. Strict creationists continue to make this argument today, in spite of the many fossils intermediate between major taxonomic groups that have been discovered.

The first, and in many ways still the most spectacular, discovery of a fossil intermediate between major groups was *Archaeopteryx* (“*primitive bird*”), found in 1861, just two years after Darwin published. It was discovered in the famous quarry near Solenhofen, Germany, where a fine-grained limestone was mined because it was especially good for making lithographic plates. The limestone is a stratum of the Jurassic period that is about 140 million years old. The first birdlike fossil discovered was only the impression of a single feather. It was in strata far older than any known at the time to contain birds. How wonderful and frustrating—birds must have been present, since no other class of vertebrates has feathers, but what were they like? A few months later an entire specimen was discovered and was eventually bought by the British Museum. A total of seven specimens of *Archaeopteryx* are now known.

Careful study showed that *Archaeopteryx*, which was about the size of a pigeon, had a blend of reptile and bird characteristics. Its feathers are characteristic of birds, but its jaws and teeth are reptilian; modern birds do not have teeth. The upper portion of the skull is birdlike, and the front appendages are modified as wings. But unlike modern birds, three of the fingers are not fused, and they end in claws. A birdlike wishbone is present, but the sternum has no keel, a structure characteristic of most modern birds. The absence of a keel suggests that *Archaeopteryx* was a glider, not a true flier, because the keel in modern birds serves as the attachment for the strong muscles that flap the wings. And finally, *Archaeopteryx* had a very reptilian long tail.

Thus, *Archaeopteryx* is a mosaic of reptilian and avian characteristics and as intermediate between two major groups of organisms as one could desire. However, a choice regarding its category of classification had to be made for practical reasons. Because its body was covered with feathers, taxonomists decided that *Archaeopteryx* should be classified as a bird. Had only the skeleton been preserved, *Archaeopteryx* would have been identified as a reptile. Indeed, exactly that error had been made before *Archaeopteryx* was discovered. Some fossils consisting only of bones were so similar to small dinosaurlike reptiles that they were classified as reptiles. Later, it was found that those bones were identical to the ones in the complete skeleton of *Archaeopteryx* and were, therefore, fossils of *Archaeopteryx*. The bodies of these individuals must have undergone considerable decay before being covered by silt, so that only bones—no feathers—remained to be fossilized. This is a dramatic illustration of the intermediate nature of *Archaeopteryx*: without feathers, they could be considered reptiles; with feathers, they are considered birds.

Evolution is a seamless strand devoid of major changes in any brief span of time. What appear as major changes are in fact artifacts in our knowledge usually due to the absence of an adequate series of fossils. When relatively few fossils of the reptiles that were the putative ancestors of birds were known, the structural gap between reptiles and *Archaeopteryx* seemed pronounced—but only because of the paucity of data. Those data are now becoming available. Numerous fossils are being discovered that are closing the gap between advanced reptiles and the earliest birds. It is clear that the many characteristics that distinguish reptiles from birds did not change all at once. One would predict, therefore, that there must have been a long period during which the evolving population could not be called either reptile or bird. At no point could one draw a line and say that every generation up to this point is reptilian and that all subsequent generations are birds. For that to be possible one would have to choose one parental generation to be labeled "*reptiles*" and call its children "*birds.*" Paleontologists today are digging up the evidence for the transition period and talking about dinosaurs with feathers.

About a decade after the discovery of Archaeopteryx, two very different species of toothed birds were discovered in deposits from the *Cretaceous period* in the western United States. One was named Hesperornis ("*place of sunset*," or "*western*," plus "*bird*") and the other *Ichthyornis* ("fish bird"). *Hesperornis* was very large, standing about three feet tall. The wings, however, were small, and the form of its body suggested that it was a diving bird, similar to the loons of today. *Ichthyornis* was gull-like in size and possibly in habits. In contrast to *Archaeopteryx*, it had a well-developed keel on the breastbone that suggested it was a strong flyer. Both genera had teeth, which strengthened the evidence that birds had descended from reptiles.

These Cretaceous birds were collected by Othniel Charles Marsh of Yale University and were described by him in a classic of paleontology, Odontornithes: Monograph on the Extinct Toothed Birds of North America (1880). As is usually the case, there are fascinating stories behind these early collecting expeditions in the American West. Settlements were few west of the Mississippi River, and travels beyond could be difficult and dangerous. Some of the flavour of what it meant to be a paleontologist in those times is provided by this description in Marsh's introduction.

The first Bird fossil discovered in this region was the lower end of the tibia of *Hesperornis*, found by the writer in December, 1870, near the Smoky Hill River in Western Kansas. Specimens belonging to another genus of the *Odontornithes* were discovered on the same expedition. The extreme cold, and danger from hostile Indians, rendered a careful exploration at that time impossible.

In June of the following year, the writer again visited the region, with a larger party, and a stronger escort of United States troops, and was rewarded by the discovery of the skeleton which forms the type of *Hesperornis regalis*.

Although the fossils obtained during two months of exploration were important, the results of this trip did not equal our expectations, owing in part to the extreme heat (110 to 120 Fahrenheit, in the shade) which, causing sunstroke and fever, weakened and discouraged guides and explorers alike.

The primitive birds found in Europe and America provided the first good evidence for the intermediates between major vertebrate groups that the theory of evolution requires. Today we have many more examples. Fossils are now known that are intermediate in their characteristics between fishes and primitive amphibians and between amphibians and primitive reptiles. One of the better-documented cases is that of the fossils of the mammal-like reptiles of the *Triassic period*.

Can intermediate types be considered ancestors? For instance, were *Archaeopteryx* and the other extinct fossil birds the ancestors of modern birds? Many people in the late nineteenth century jumped to that conclusion. However, today one would be hard pressed to find a paleontologist willing to say so. Rather, researchers today are careful to recognize the limits, of their methods and only say that there were probably many different species of birds living when *Archaeopteryx* was alive, and one or more of these must have been the ancestors of later types. It is also quite possible that there is no present-day bird for which *Archaeopteryx* was a direct ancestor.

The fossil record for the vertebrates is now adequate to reveal the broad outlines of the evolution of the major groups: fish, amphibians, reptiles, birds, and mammals. A great deal is also known about evolution within every group except the birds. The data are especially good for mammals and reptiles. Mammals have the advantage of a fairly recent origin and so a shorter period for their fossil record to be destroyed, and their skeletons and teeth provide excellent material for determining lineages. Teeth are especially useful because they vary greatly among the different groups of mammals. Some paleontologists have said, only partly in jest, that it is possible to determine the evolutionary history of mammals from their fossil teeth alone. The teeth of fish, amphibians, and reptiles are less useful because they are fairly uniform, and modern birds are toothless.

The mammals also provide some good evidence regarding evolution within families, the horses being a classic example. In 1874 Marsh published an early version of the story of the evolution of the horse, based on fossils collected on his western expeditions:

> The animals of this group which lived in this country during . . . the *Tertiary period* were especially numerous in the Rocky Mountain regions, and their remains are well preserved in the old lake basins which then covered so much of that country. The most ancient of these lakes—which extended over a considerable part of the present territories of Wyoming and Utah—remained so long in Eocene times that the mud and sand, slowly deposited in it, accumulated to more than a mile in vertical thickness. In these deposits, vast numbers of tropical animals were entombed, and here the oldest equine remains occur.

But Marsh found more than the earliest fossil horses. He found four fossil species, each restricted to a different layer of Tertiary rocks but together spanning almost the entire *Tertiary period*. Today geologists divide the Tertiary period into five *epochs*, from *Paleocene* through *Pliocene*. There follows the very brief *Quarternary period*, including the *Pleistocene* and *Recent epochs*, the latter a mere ten thousand years. It was in the Recent epoch that human beings became a dominant force on Earth.

The first fossil horse Marsh discovered was *Orohippus*, from the *Middle Eocene*. Several years later, however, he found *Hyracotherium* from the *Early Eocene*, which remains the most ancient known genus of the horse family. Next in order of descent was *Miohippus* from the *Miocene*, and the youngest fossil was *Hipparion*, from the *Pliocene*. The modern horses belong to the genus Equus.

Thus Marsh could consider five genera that gave a view of the history of horses over the last 50 million years. To be sure, an average of one ancient horse for every 10 million years cannot be considered a very complete record, but at least it was a beginning. The differences in the anatomy of these horses were not random but seemed to exhibit trends that allowed one to understand how the *Hyracotherium* structures gradually changed into those of other genera to finally produce the modern horses of the genus *Equus*. The two *Eocene* genera, *Hyracotherium* and *Orohippus*, were very small animals, about the size of a fox or small dog. The size of horses increased during the *Tertiary period*, *Miohippus* (from the *Miocene*) and *Hipparion* (from the *Pliocene*), each being progressively larger. In fact, *Hipparion* was as large as some of the smaller breeds of contemporary horses.

Of exceptional interest is the evolution of the structure of the horses' front legs. The more primitive *Paleocene* mammals from which horses presumably evolved had five digits on each appendage, as is the case with our hand and foot. The *Eocene* horses, however, had four toes on the front feet and three toes on the hind feet, each with a tiny hoof. Counting the innermost toe of a five-digit appendage (like our thumb or big toe) as number I, the front legs had lost number I and only numbers II, III, IV, and V remained. The hind foot had lost toes number I and V. The discovery of little horses with three or four toes, each with a little hoof, proved fascinating not only to paleontologists but to the general public as well. Both *Miocene Miohippus* and *Hipparion* in the *Pliocene* had three toes on the front feet, but the side toes of *Hipparion* were smaller than the other toes.

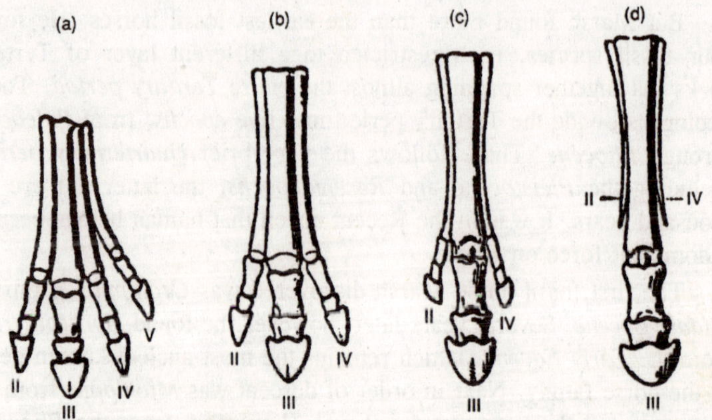

(a) Orohippus (Eocene); (b) Miohippus (Miocene); (c) Hipparion (Pliocene); (d) Equus (Quaternary).

Fig. 8.1. Marsh's original drawing illustrating the evolution of the horse's feet begins with the four-toed genus of the Eocene, Orohippus, and continues with the Miocene (Miohippus) and Pliocene (Hipparion) genera.

Modern horses have a single toe, number III, on each foot, and the vestigial remains of toes II and IV are present as the splint bones, as shown in Marsh's illustration. All horses of today walk around on what were once the third toes of their ancestors. Even before these fossil horses were discovered, comparative anatomists had concluded that the modern horse had a single functional toe and the splint bones were vestigial toes. The discovery of fossil horses validated their hypothesis.

There were many other changes in the skeleton, including a great increase in size from the fox-size dimensions of *Orohippus* to *Equus* today. The lengthening of the limb bones permitted increased running speed. The neck and the front part of the skull—the area in front of the eyes—became elongated.

Horses remain one of the most interesting examples of the evolution of a vertebrate family. In spite of the fact that today North America has no native horses, it was the center of horse evolution. A worldwide total of about 34 genera of fossil horses have been discovered, and all became extinct except one, *Equus*, which includes horses, zebras, asses (donkeys), and onagers. Although having 34 genera to represent about 50 million years of evolution might not seem like a paleontological cornucopia, the relative abundance of individual fossils has provided considerable information about the habits and distribution of the members of the horse family.

The structure of the teeth suggests that the horses that lived before the *Miocene* were browsers (that is, they ate the leaves of trees) rather than grazers (which eat grasses and other low vegetation). This change occurred over a period of a few million years across many different genera of horses and can be attributed to a worldwide climate change. The gradual cooling and drying of the climate during the *Miocene* led to a decrease of forests and an increase in grasslands, which prompted the horses to switch from browsers to grazers worldwide. That introduced a new problem. The silicon in grasses is a very abrasive substance that causes wear on the teeth. Correlated with the new environmental challenge, natural selection promoted a change in the teeth. The adult teeth of other mammals, such as human beings, do not grow after they appear, but those of horses and many other grazing animals continue to grow throughout life; they thus do not become worn down to the gums but remain full-sized.

Nearly all the *Miocene* genera of horses became extinct, and not many were left during the *Pliocene*; but *Equus* appeared in North

America, and several genera migrated to South America. *Equus* also migrated to the Old World, and various species evolved in Europe, Asia, and Africa. All New World horses were extinct by about 11,000 years ago, when the *Pleistocene Ice Age* ended. The cause is unknown, but it was part of a mass extinction that saw the death of many other large mammals, an estimated 186 species in all. The New World mammoths, mastodons, camels, giant sloths, giant bears, and sabertooth cats vanished. Horses survived only in the Old World. Today's "*wild*" horses of North and South America are descendants of domesticated European horses introduced by the Spanish in the sixteenth century.

The horse story became known only a decade after the discovery of *Archaeopteryx*, and it suggested the same general process—evolution. It is important to emphasize, once again, the nature of the evidence. The sequence in geological time of *Hyracotherium*, *Orohippus*, *Miohippus*, *Hipparion*, and *Equus* shows a series of intermediate stages that link the *Eocene horses*, *Hyracotherium*, with the living species of *Equus*. But a careful paleontologist would not say, for instance, that *Miohippus* was the ancestor of *Hipparion*; only that it could have been. The possibility that another unknown genus very similar to *Hipparion* was the true ancestor cannot be dismissed. The concept of evolution is supported by the discovery of intermediate fossils, not "*true*" ancestors. Thus, the hypothesis would be that if Equus evolved from some Eocene horse through a long line of intermediates, the paleontological record should contain fossils intermediate between *Hyracotherium* and *Equus*. Diligent and lucky digging might reveal these intermediates. Marsh was both diligent and lucky. The evidence for a direct ancestor to a descendant would require an enormous number of fossil specimens from closely similar times.

The story of horse evolution was brought to the American public by the outstanding English naturalist and vigorous defender of Darwinism, Thomas Henry Huxley. On September 20, 1876, two years after Marsh had published his results, Huxley gave a series of public lectures in New York City. He had visited Marsh at Yale University and learned firsthand the story of fossil horses in the American West. His third lecture, "The *Demonstrative Evidence of Evolution*," was about the evolution of horses' toes.

An inductive hypothesis is said to be demonstrative when the facts are shown to be in entire accordance with it. If that is not scientific proof, there are no merely inductive conclusions which can be said to be proved. And the doctrine of evolution, at the present time, rests upon exactly as secure a foundation as the Copernican theory of the motions of the heavenly bodies did at the time of its promulgation. Its logical basis is precisely of the same character—the coincidence of the observed facts with theoretical requirements.

The only way of escape, if it be a way of escape, from the conclusions, is the supposition that all these different equine forms have been created separately at separate epochs of time; that of such an hypothesis as this there neither is, nor can be, any scientific evidence; and, assuredly, there is none which is supported, or pretends to be supported, by evidence or authority of any other kind. Think that the time will come when such suggestions as these, such obvious attempts to escape the force of demonstration, will be put upon the same footing as the suppositions made by some writers, who are, not completely extinct at present, that the fossils are mere simulacra, are no indications of the former existence of the animals to which they seem to belong; but that they are either sports of Nature [what we would now call mutations], or special creations, intended.

In fact, the whole evidence is in favour of evolution, and there is none against it. Although perfectly well aware of the seeming difficulties which have been built up upon what appears to the uninformed to be a solid foundation.

Huxley's analysis was accurate when he spoke to that New York audience in 1876 and remains so today. Surely that would please him. But he would be discouraged to find that the same sorts of criticisms of evolution current in his time are still accepted today by the uninformed and by those who reject the statements of science if they conflict with religious beliefs.

Radioactive Dating

What if Marsh had set up camp in a promising area having cliffs with thick layers of stratified rocks; each of four of his assistants had returned with a different fossil horse that would in time be named *Hyracotherium*, *Orohippus*, *Miohippus*, and *Hipparion*; and close examination of the fossils had revealed that the four genera seemed to form a series of increasing size and complexity. Could one reasonably conclude that these four ancient horses were an evolutionary series? Not unless it was found that the four fossils came from four different geological strata that matched in order, from oldest to youngest, the order of the fossils from smallest and simplest to largest and most complex. Older and younger are adequate for assessing relative age, but clearly an accurate method of determining exact geological ages was needed. Various methods were suggested for determining absolute ages but no acceptable method was found. That sad state of affairs is illustrated by the range of dates proposed for the beginning of the *Cambrian period*–when the first chordates came on the scene–ranging from 70 million years to 6,000 million years ago.

As is so often the case in scientific research, a solution in one field comes from discoveries in another field. In the case of how to tell geological time the solution came from physics. It was radioactivity that provided a method for geologists to obtain confirmable evidence for the true age of a fossil. Before the 1890s atoms were defined as the ultimate particles of matter–the smallest physical objects in existence and indivisible (the word atom is from a Greek word meaning "*indivisible*"). Atoms were known to differ from element to element, but all atoms of a given element were thought to be identical. This was a reasonable deduction, since chemical reactions between the same elements always produced the same product or products. One oxygen and two hydrogen atoms always combine to form water, and for all intents and purposes water is water.

During the 1890s three scientists in France–A. H. Becquerel, Polish-born Marie Curie, and her husband, Pierre Curie–embarked on a series of experiments that would revolutionize physics, chemistry, biology, and to some extent geology. Physicists were beginning to suspect that atoms were composed of smaller particles, that they were neither indivisible nor constant, and that elements could change into other elements under certain conditions. Such changeable elements are said to be radioactive: in the process of changing from one element to another, they emit subatomic particles. For example, radioactive uranium goes through a series of about a dozen steps in which it becomes thorium, protoactinium, radon, polonium, bismuth, and other elements until it finally ends up as lead. A striking feature of this radioactive decay–and the one that makes it useful to geologists–is that the rate of this radioactive decay is constant and cannot be influenced by any known physical condition, such as temperature or pressure. This is why an atomic clock, which measures the rate of disintegration of radioactive elements, is amazingly accurate.

The time required for half of the atoms in a radioactive sample to reach a nonradioactive state is called the element's half-life. In the case of *uranium*, the half-life is the interval required for half of a sample of *uranium* to become lead; scientists determined that this takes roughly 4.5 billion years. Thus the ratio of uranium to lead in a sample can be used to determine when the reactions began. For example, if a sample of rock had equal amounts of *uranium* and *lead*, that rock would have been formed 4.5 billion years ago. Great care must be taken in selecting the sample of rock to be analyzed. A most important requirement is that no loss of the breakdown products of *uranium* should have occurred; they must be trapped in the rock and measured. If any portion is lost, the estimate will be incorrect.

How can one be sure the lead in the sample of rock being tested came from the breakdown of uranium and was not just a normal part of the rock? The answer involves the notion of isotopes. By the time the uranium method for dating rocks was being developed, it had been discovered that all of the atoms of an element need not be identical; they could differ in the number of subatomic particles of which they are composed. These different versions were called isotopes. There are several isotopes of uranium; the one used for age determination is $_{238}U$, the 238 being its atomic mass. When this uranium isotope disintegrates, it ends as a rare isotope of lead known as $_{206}Pb$. The common isotope of lead, which is present in many rocks, is $_{207}Pb$. The uranium method for determining the age of a rock cannot work unless one can distinguish $_{206}Pb$ from $_{207}Pb$. Methods are available to do this with a high degree of accuracy. Therefore age can be determined by the ratio of $_{238}U$ to $_{206}Pb$.

Today many different techniques to date materials are based on radioactivity. The radioactive isotope most often used in dating recently living objects is carbon 14 ($_{14}C$). It is especially useful for dates in human prehistory because it has a very short half-life—5,700 years. In carefully prepared samples it can be used to date carbon containing material, such as wood, that is as old as 70,000 years. Different isotopes can be used to date materials from longer spans of geological time—for example, the conversion of potassium to argon (half-life of 1.210_9) and rubidium to strontium (half-life of 50 10_9). Whenever two methods are used to date the same sample and they give the same answer, we can be more confident that the answer is correct.

With radiometric methods available for accurately dating sedimentary rocks, geologists could estimate the ages of strata with increasing exactitude. The great age of the Earth and the vastness of the time that life has been evolving came as a great shock to some. The age at which our solar system, including the Earth, began to form from cosmic dust is now estimated at about 4.5 billion years ago—a far cry from 4004 b.c., Bishop James Ussher's estimate of the year

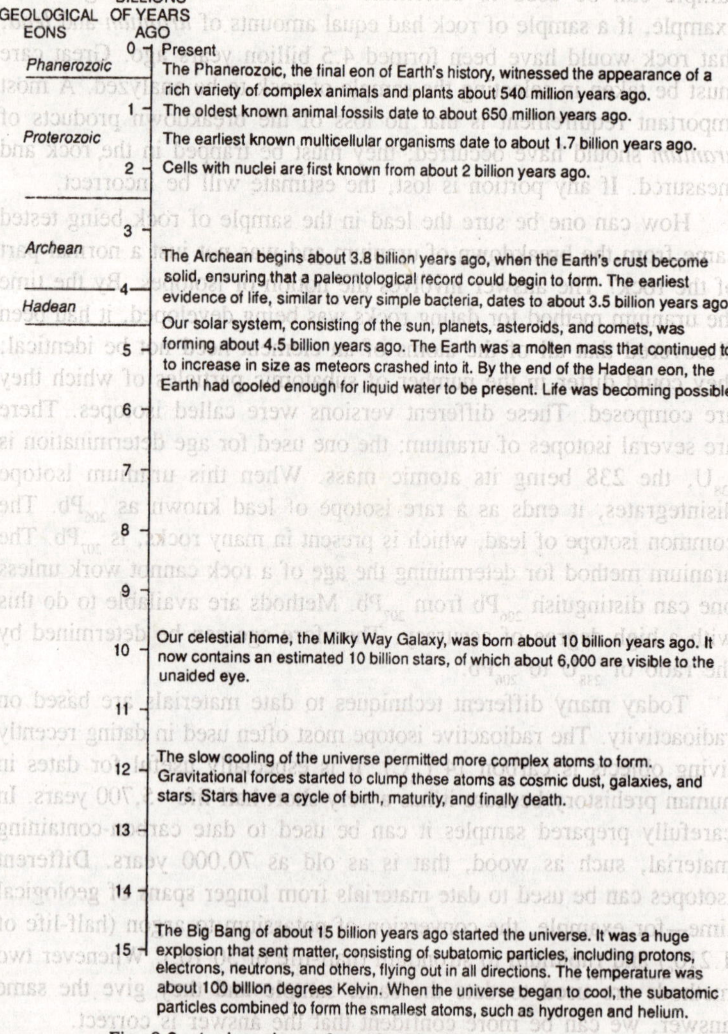

Fig. 8.2. Ancient times: from the Big Bang to the present.

God created the heavens and the earth. This new value is based mainly on radioactive dating of meteorites, which formed at the same time as the sun and planets. Geologists do not base estimates of the age of the Earth on radioactive dating of Earth's own rocks, because at its beginning the Earth was a hot, molten mass, and the oldest rocks could have been formed only after the Earth had cooled. The oldest rocks known on Earth have been dated to about 3.8 billion years old.

One can learn from these very old rocks something about the composition of the Earth's crust and its atmosphere at that time, and with luck, possibly find evidence of any life that might have existed then.

Life Over Time

Throughout the nineteenth and early twentieth centuries, there was no good fossil evidence for life before the *Cambrian period*, which began about 540 million years ago. It seemed that in the Cambrian period an impressive variety of animals had suddenly appeared—complex representatives of most of the major phyla that are still alive today. How could one account for the sudden appearance—on a geological time scale—of such relatively complex animals as shellfish, starfish, corals, and crustaceans? One hint of the answer is that all of the *Cambrian* animals known at the time had shells or other hard structures. Soft body parts such as internal organs very rarely fossilize, so animals in the *Precambrian*, which had not yet evolved complex shells, would not have been so readily preserved. The animals of the *Cambrian* only appeared to have suddenly shown up; there is now abundant evidence that these complex animals were the results of a long period of evolution in the *Precambrian*.

In recent decades special techniques have been developed for searching very ancient sedimentary rocks for evidence of soft-bodied animals and even simple bacteria. Before these techniques were available, it was far more exciting to search for the noble dinosaurs than for microscopic creatures with the simplest anatomy. But by the middle of the twentieth century the general patterns of animal evolution were fairly well known, and the question of what happened in those vast eons before the *Cambrian* became a prime topic for research. As a result of these investigations, the date for the first evidence of life has been pushed back to about 3.5 billion years ago, shortly after the first rocks formed on Earth. Life at the beginning consisted of primitive bacteria, some of which cannot be distinguished from the prokaryotes living today.

The *prokaryotes* were the only forms of life in the interval from about 3.5 to about 2 billion years ago when eukaryotes came on the scene. Both the prokaryotes and the eukaryotes have the same hereditary material, DNA, but in the prokaryotes it lies free in the cell. In eukaryotes most of the DNA is enclosed in a vesicle called the *nucleus*. (DNA is also found in tiny organelles within cells, called the *mitochondria*, and in the chlorophyll of green plants.) From 2 billion to about 650 million years ago there is no generally accepted fossil

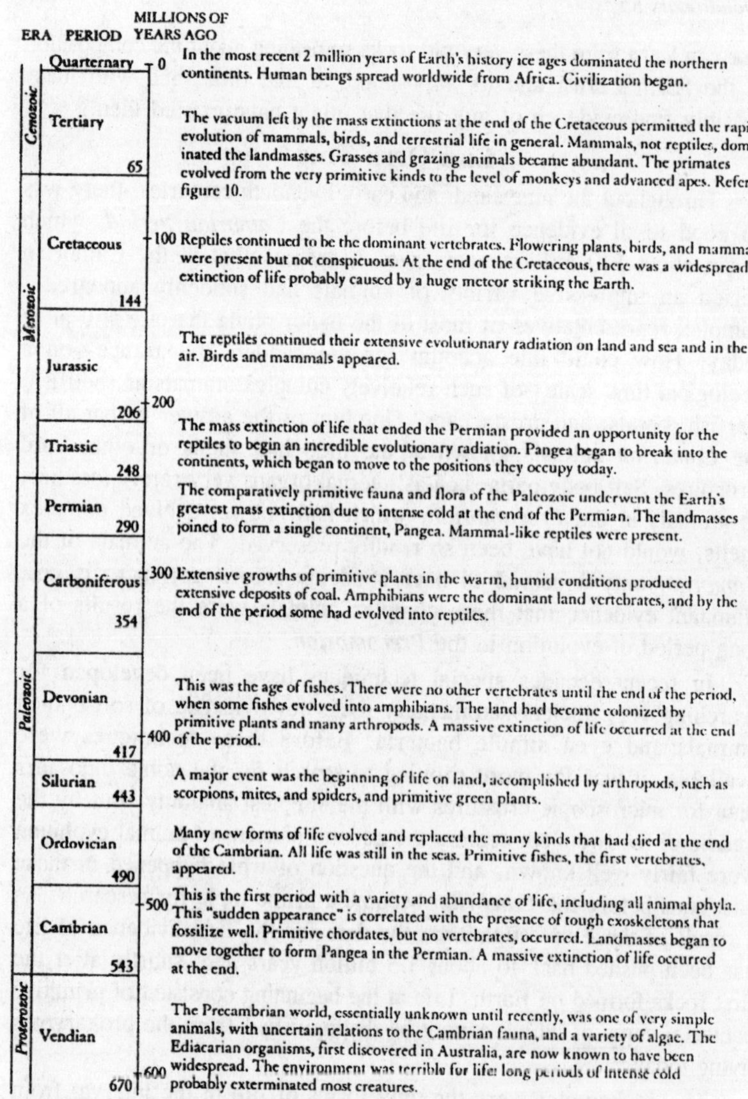

Fig. 8.3. Ancient times: the past 600 million years.

evidence for animals that are composed of many eukaryotic cells—the multicelled animals, or metazoans. Recent discoveries have found metazoans beginning about 650 million years ago. By the onset of the *Cambrian*, 540 million years ago, metazoans appeared in abundance. The Precambrian strata are now being intensively searched for evidence

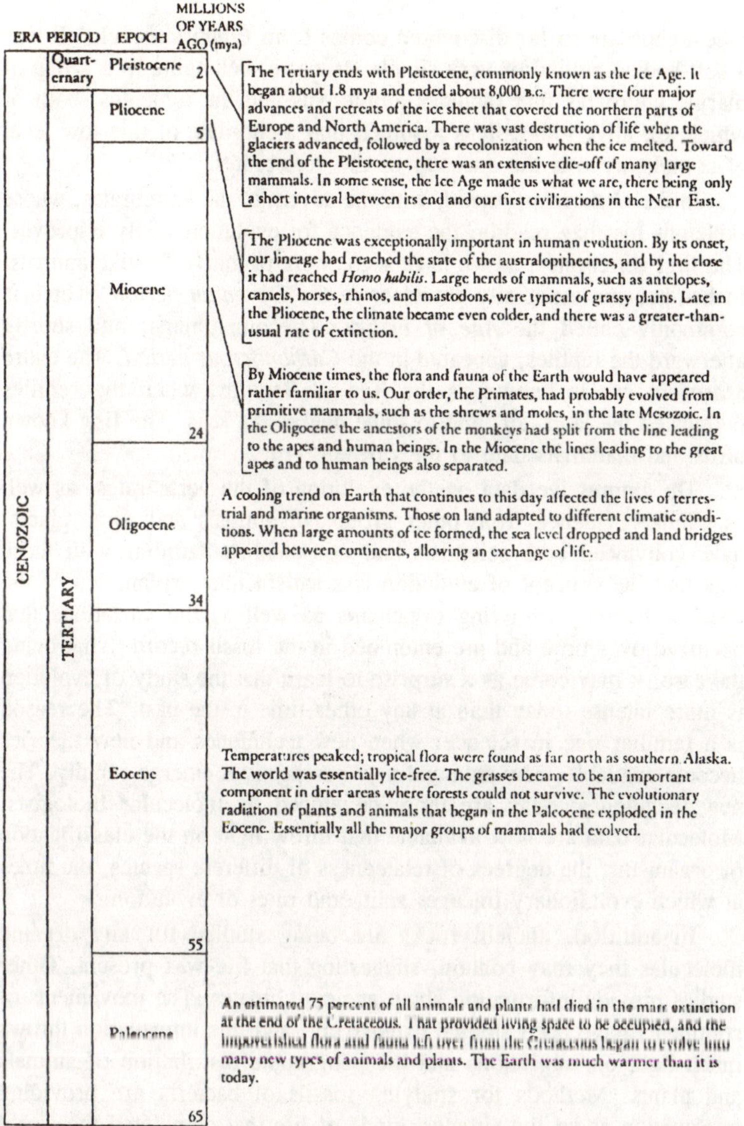

Fig. 8.4. Ancient times: the past 65 million years.

of soft-bodied multicelled animals and with considerable success. The oldest group of soft-bodied metazoans to be discovered to date is in Australia, but others are turning up in other parts of the world. Looking at our own phylum, the Chordata, the fossil record shows that the oldest chordate so far discovered comes from *Cambrian* rocks. It was a soft-bodied animal of very simple structure, belonging to a group of marine chordates that includes amphioxus and the tunicates, both of which are alive today. It is assumed that a chordate of this low level of complexity was the ancestor of the vertebrates.

In the *Ordovician period*, with the advent of the vertebrates, whose skeletons fossilize readily, the evidence for evolution vastly improves. The first vertebrates, as we have seen, were primitive fishlike animals. Fishes became especially abundant in the *Devonian period*, which is commonly called the *Age of Fishes*. The amphibians, and shortly afterward the reptiles, appeared in the *Carboniferous period*. The entire *Mesozoic era* is known as the *Age of Reptiles*, when the reptiles dominated the land, air, and to some degree the seas. The first known birds and mammals date to the *Jurassic era*.

The impressive data on the evolution of the vertebrates, as well as similar evidence from other groups of animals and some plants, have convinced both scientists and nonscientists familiar with those data that the concept of evolution is a satisfactory explanation of the present diversity of living organisms as well as the variations that occurred over time and are entombed in the fossil record. That being the case, it may come as a surprise to learn that the study of evolution is more intense today than at any other time in the past. The reason is a familiar one in science: when new techniques and new theories become available, new data and new perspectives emerge rapidly. The new techniques today are those developed by molecular biologists. Molecular data are now available that throw light on the classification of organisms, the degrees of relatedness of different species, the times at which evolutionary lineages split, and rates of evolution.

In addition, ancient rocks are being studied for any organic molecules they may contain, suggesting that life was present. Other studies provide information about ancient climates. The movements of continents have been studied extensively, and this information throws much light on migrations and the geographic distribution of animals and plants. Methods for studying fossils of bacteria are providing information about the simplest kinds of life that dominated the world for billions of years—and in some ways still do. The *Precambrian*—once a dark period in our knowledge of the past—is finally opening up to scientific scrutiny. Not only can new questions about evolution be asked and answered, but many solutions to age-old puzzles about the secrets of life are being unlocked from the Earth's rocks for the first time.

9

TRAIL ON EVOLUTION

In the nineteenth century the United States witnessed a great ferment in Christianity that resulted in many schisms that persisted into the twentieth century. The religious climate was very different from that in most nations of Western Europe, which had established or dominant religions, such as the Episcopal (Anglican) Church in England and the Roman Catholic Church in France, Italy, and Spain. In the United States, some of the sects took an extreme fundamentalist position and insisted on the inerrancy of the Bible. One of the founding fathers of American fundamentalism was the evangelist Dwight L. Moody, whose position was that a line should be drawn between the church and the world and every Christian should get both feet out of the world. Among the sins of the world to be avoided were activities on Sunday such as sports, entertainment, reading the newspaper, attending the theater, and on all days, dancing, card playing, liquor drinking, pandering to the lusts of the flesh, and atheistic teachings such as evolution.

Fundamentalists were vehemently opposed to the theory of evolution, since *Genesis* provided a very different account for the origin and diversity of life. In previous centuries many theologians had interpreted the *Bible* to say that the *Earth* is flat, that the *Earth* is the center of the universe, and that the sun rotates around the *Earth*. The voyage of Magellan in 1522 had provided convincing evidence that the Earth is not flat, but news seems to travel slowly, for in 1922 a school-teacher in Kentucky was fired for teaching that the Earth is round (Ginger 1958). In 1543, just a few years after Magellan's voyage, Copernicus presented evidence that the sun, not the *Earth*, is the center of our solar system. His views, too, were eventually accepted.

As time went on, more and more religious leaders relaxed their demands for inerrancy of the Bible relative to the discoveries that the Earth is neither flat nor at the center of the universe. The clergy were able to accept these apparent violations of Holy Writ and so could their congregations. This has not happened with evolution, however. Well into the twentieth century, William Jennings Bryan, a famous Democratic politician and even more famous orator, proclaimed that "the evolutionary hypothesis is the only thing that has seriously menaced religion since the birth of Christ and it menaces all other religions as well as the Christian religion, and civilization as well as religion" (1923).

Scopes Trial

A major challenge to the dominance of creationism and the rejection of evolution occurred in Dayton, Tennessee, in 1925. John Thomas Scopes was tried for teaching Darwinism in his high school biology class—in defiance of an act recently passed by the General Assembly of the State of Tennessee. The trial brought the evolution-creationism controversy to the nation's attention and emphasized the polar positions of the two sides.

On January 28, 1925, the lower house of the Tennessee legislature passed a bill introduced by John Washington Butler. The critical paragraph read as follows. "Section 1. Be it enacted by the General Assembly of the State of Tennessee, that it shall be unlawful for any teacher in any of the Universities, Normals and all other public schools of the State, which are supported in whole or in part by the public school funds of the State, to teach any theory that denies the story of the Divine creation of man as taught in the Bible, and to teach instead that man has descended from a lower order of animals." (Unless noted otherwise, the quotations are from the transcript of the Scopes trial, 1925.)

Butler was no Bible-thumping firebrand but a kind, friendly gentleman who had become concerned when a young woman of his community had attended a university and returned home believing in evolution but no longer in God. Butler worried that a similar fate might befall his children, and he decided to do something effective: he ran for the state legislature and was elected. Part of his platform was the need to prohibit the teaching of evolution because it might corrupt young people. He introduced a bill to do just that, and by coincidence, the well-known politician, statesman, and fundamentalist William

Jennings Bryan delivered a lecture in Nashville the very night following the introduction of Butler's bill. His title was "Is the *Bible* True?" Bryan argued strongly that it was. Copies of his lecture were distributed to the members of the legislature and may have swayed some of them. In any case, Butler's bill passed the upper house and was signed by Governor Austin Peay, who expressed the opinion that the bill was a distinct protest against an irreligious tendency to exalt so-called science and to deny the *Bible*. He went on to say that the tendency was fundamentally wrong and fatally mischievous in its effects on children, institutions, and the country.

There was no organized opposition to the Butler Act, even though some politicians and others may have thought it an error to enact such a law. For office seekers it would have been folly to deny the truth of the Bible, which was the basis of the religious beliefs of most voters in Tennessee. It was the New York City-based American Civil Liberties Union that decided to test the validity of the Butler Act. The ACLU arranged for Scopes, a 24-year-old high school biology teacher and graduate from the University of Kentucky, to be charged with teaching evolution in defiance of the Butler Act. The ACLU hoped to use the trial to test the constitutionality of the law. The position of the defense was that it was not proper for any legislature to pass such a law and that prohibiting the teaching of Darwinism was analogous to prohibiting the teaching of heliocentrism—that the sun is the center of the universe. Clarence Darrow, a prominent trial lawyer, became the chief lawyer for the defense. Bryan offered his services to the prosecution, saying that the state could not afford to have a system of education that destroys the religious faith of children.

The trial began in Dayton on Friday, July 10, 1925. The small town's population was greatly swollen by numerous visitors from out of state, more than a hundred newspaper reporters, local people who came in from the surrounding hills and farms, and merchants to supply the needs of all for food, trinkets, toy monkeys (representing ancestors), and *Bibles*. The Scopes trial was seen as big news. It defined the antagonists and the concerns about teaching evolution that are still with us at the beginning of the third millennium. The trial was a contrived event initiated by the ACLU and supported by those modernists who sought to prevent religious points of view from deciding what was to be taught in the public schools. On the opposing side, the defenders of the Butler Act were determined to maintain their way of life and system of belief.

Presiding Judge John T. Raulston called the court to order. He read the Butler Act and then the first chapter of Genesis. Interestingly enough, the Bible was identified as the "St. James version rather than the King James version". The jury was instructed not to pass judgment on the constitutionality of the Butler Act but simply to decide whether or not Scopes was guilty of teaching evolution. The main lawyer for the prosecution, Attorney General A. T. Stewart, read that section of the state constitution relative to freedom of religion, which said that no preference should be given to any religious establishment or mode of worship. Darrow, the defense lawyer, suggested that the Butler Act gave clear preference to the *Bible* and asked why not, in that case, allow the Koran as well. The attorney general answered that the Bible was preferred because they were not living in a heathen country.

On the second day of the trial Darrow made a lengthy statement indicating what the defense hoped to accomplish, namely, to have the Butler Act declared unconstitutional. Darrow held Bryan guilty for such a pernicious law's being passed in the first place: he had a long history of antievolution activity and had encouraged, without success, the passage of an antievolution law in Florida. This excerpt and others that follow will give the flavour of Darrow's remarks.

Mr. Bancroft [a historian] wrote this sentence, which is true: "That it is all right to preserve freedom in constitutions, but when the spirit of freedom has fled, from the hearts of the people, then its matter is easily sacrificed under law." And so it is, unless there is left enough of the spirit of freedom in the state of Tennessee, and in the United States, there is not a single line in any constitution that can withstand bigotry and ignorance when it seeks to destroy the rights of the individual; and bigotry and ignorance are ever active. Here we find today as brazen and as bold an attempt to destroy learning as was ever made in the middle ages, and the only difference is we have not provided that they shall be burned at the stake, but there is time for that.

Darrow then argued that the law was invalid because it was not drawn up properly. Further, it was wrong to specify the Bible as the divine book and ignore the Koran, the Book of Mormon, the beliefs of Confucius or Buddha, or even Emerson's essays on transcendentalism. Darrow was aware of the "*higher criticism*" of the *Bible* because he informed the court that there are two conflicting accounts of creation in the first two chapters of Genesis. The Butler Act had not said which account of creation was at issue. About why Scopes was being charged Darrow said, he knew he was there because the fundamentalists were after everybody who thought, because ignorance and bigotry were rampant. Darrow ended his speech on this second day of the trial as follows:

I will tell you what is going to happen, and I do not pretend to be a prophet, but I do not need to be a prophet to know. Your honor knows the fires that have been lighted in America to kindle religious bigotry and hate. You can take judicial notice of them if you cannot of anything else. You know that there is no suspicion which possesses the minds of men like bigotry and ignorance and hatred. . . .

If today you can take a thing like evolution and make it a crime to teach it in the public school, tomorrow you can make it a crime to teach it in the private schools, and the next year you can make it a crime to teach it to the hustings or in the church. At the next session you may ban books and the newspapers. Soon you may set Catholic against Protestant and Protestant against Protestant, and try to foist your own religion upon the minds of men. If you can do one you can do the other. Ignorance and fanaticism is ever busy and needs feeding. Always it is feeding and gloating for more. Today it is the public school teachers, tomorrow the private. The next day the preachers and the lecturers, the magazines, the books, the newspapers. After a while, your honor, it is the setting of man against man and creed against creed until with flying banners and beating drums we are marching backward to the glorious ages of the sixteenth century when bigots lighted fagots to burn the men who dared to bring any intelligence and enlightenment and culture to the human mind.

Much of the third day was given over to a discussion of whether or not each session of the court should begin with a prayer, and the judge decided that having prayers would be a good thing. In a minor episode, the attorney general characterized one of the defense lawyers as "the agnostic counsel for the defense" and told another, Mr. Hays, to shut up. The next day the attorney general offered his apology, and Mr. Hays responded: "Permit me to say personally that there are two qualities I much admire in a man. One is that he is human and the other is that he is courteous. The outburst on yesterday proves that the attorney-general was human, and the apology proves that he has the courtesy of a southern gentleman".

On the fourth day Judge Raulston read a long opinion on why the motion of the defense to quash the indictment was rejected. In the afternoon the attention of the trial finally turned to Scopes. The position of the prosecution was simple: there was a law that forbade the teaching of evolution, and Scopes had done so. Even the defense agreed to that.

The position of the defense was much more complicated. It held that it was necessary not only to prove that Scopes had taught evolution but also that he had denied the theory of creation given in the Bible. But the Bible has more than one theory of creation (the P and J versions), so which one had Scopes denied? Further, the defense maintained that there is no conflict between evolution and Christianity and noted that in earlier years Bryan had held the liberal view that religion was not subject to legislation and that no science can be taught without recognizing evolution. The defense asked that the fundamentalist Bryan of today revert to the modernist Bryan of yesterday. It pointed out that Christianity had survived in spite of all the discoveries of science and that science occupied a field of learning separate and apart from that of theology. The defense mentioned vestigial organs as indicative of evolution, but no evidence was presented for the evolution of human beings from monkeys, as the prosecution had said. Further, the defense argued that the *Bible* cannot be regarded as an adequate treatment of science because of the many discoveries made since it was written. For example, "Moses never heard about steam, electricity, the telegraph, the telephone, the radio, the aeroplane, farming machinery, and Moses knew nothing about scientific thought and principles from which these vast accomplishments of the inventive genius of mankind have been produced".

Witnesses for the prosecution were then called to testify that Scopes had admitted teaching the section on evolution that was in the textbook approved by the state for use in its schools—George William Hunter's Civic Biology. Scopes said he had done so because he thought that the statute was unconstitutional. Surprisingly, the jury was excluded when Darrow and the other defense lawyers called witnesses for the defense. The first and only expert witness permitted to testify was the well-known zoologist Maynard M. Metcalf, who went over the scientific evidence for evolution. Judge Raulston did not permit the other scientists or any of the theologians present to testify for the defense. However, they were allowed to submit written statements, which were to be printed in the official record of the trial.

When Judge Raulston asked Darrow if he wished to call Scopes to the witness stand, Darrow replied that every charge against the defendant was true, and when Darrow was asked again, he replied that there was no point in doing so. Scopes never was called as a witness. Again, the strategy of the defense was to have the Butler Act declared unconstitutional, not to have Scopes found innocent. If he was found guilty, the case could be appealed to the Supreme Court of Tennessee and the Butler Act would be evaluated and, it was hoped, declared unconstitutional.

A high point was in the afternoon of the fifth day when Bryan made his first serious comments. He had come to Dayton as a nationally known orator and politician, having served three times as the Democratic Party's candidate for president. Impressive in appearance, he was a leader of the fundamentalists and a vigorous and effective antievolutionist. The people of Dayton regarded Bryan with love, respect, and reverence. They expected him to demolish the forces of evil that were attempting to corrupt their children (presumably adults were immune to this infection).

Bryan began by ridiculing evolution, and the court record shows that he was interrupted frequently by laughter from those attending the trial. He referred to a diagram in the biology textbook used by Scopes and his students showing the animal kingdom and the number of species in each major taxonomic group. Each group was in a circle with a size relative to the number of species. The species numbers were estimates, of course, but Bryan noted that all were round numbers and wondered if animals bred that way. The 3,500 mammals rated only a very small circle with hardly enough room for human beings. "There is the book! They were teaching your children that man was a mammal and so indistinguishable among the mammals that they leave him there with thirty-four hundred and ninety-nine other mammals (*laughter* and *applause*)" . Bryan also read from Darwin's Descent of Man that man evolved from Old World monkeys: "Not even the American monkeys but from Old World monkeys (*laughter*)". "Talk about putting Daniel in the lion's den? How dare these scientists put man in a little ring like that with lions and tigers and everything that is bad! Not only the evolution is possible, but the scientists possibly think of shutting man up in a little circle like that with all these animals that have an odor, that extends beyond the circumference of this circle, my friends (*extended laughter*)".

Since the defense had been prohibited from having all but one scientist testify about evolution, they asked about having expert witnesses for the *Bible*. Byran replied: "Now, your honor, when it comes to *Bible* experts, do they think they can bring them in here to instruct the members of the jury, eleven of whom are members of the church?

I submit that of the eleven members of the jury, more of the jurors are experts on what the *Bible* is than any *Bible* expert who does not subscribe to the true spiritual influences or spiritual discernments of what our Bible says (Voice in audience, '*Amen*!')".

Bryan also sought to show that some distinguished scientists had doubts about evolution. He quoted from a lecture given in Toronto by the prominent British geneticist William Bateson to prove this point. The defense countered this point with a statement by one of the scientists who had submitted in writing what he had intended to say. That scientist had written to Bateson to ask his views. Bateson replied that he had looked through his Toronto address and found nothing that could be construed as expressing doubt regarding the main fact of evolution. He then expressed the opinion that the campaign against the teaching of evolution was a terrible example of the way in which truth could be perverted by ignorant people.

Dudley Field Malone, one of the defense lawyers, answered Bryan at length. He pointed out the difference between theological and scientific ways of thinking: "The main difference between the theological mind and the scientific mind is that the theological mind is closed, because that is what is revealed and is settled. But the scientist says no, the Bible is the book of revealed religion, with rules of conduct, and with aspirations—that is the *Bible*. The scientist says, take the *Bible* as guide, as an inspiration, as a set of philosophies and preachments". Malone continued:

There is never a duel with the truth. The truth always wins and we are not afraid of it. The truth is no coward. The truth does not need the law. The truth does not need the forces of government. The truth does not need Mr. Bryan. The truth is imperishable, eternal and immortal and needs no human agency to support it. We are ready to tell the truth as we understand it. . . . We feel we stand with progress. We feel we stand with science. We feel we stand with intelligence. We feel we stand with fundamental freedom in America. . . . We ask your honor to admit the evidence as a matter of correct law, as a matter of sound procedure and as a matter of justice to the defense in this case (profound and continued applause).

Judge Raulston was not swayed, and the attorney general claimed that evolution would be the end of the *Bible*:

I say, bar the door and do not allow science to enter. That would deprive us of all the hope we have in the future to come. And I say it without any bitterness. I am not trying to say it in the spirit of bitterness to a man over there [Darrow], it is in my view, I am sincere about it. Mr. Darrow says he is an agnostic. He is the greatest criminal lawyer in America today. His courtesy is noticeable—his ability is known—and it is a shame, in my mind, in the sight of a great God, that a mentality like his has strayed so from the natural goal it should follow—great God, the good that a man of his ability could have done if he had aligned himself with the forces of right instead of aligning himself with that which strikes its fangs at the very bosom of Christianity.

Yes, discard that theory [Divine Creation] of the Bible—throw it away, and let scientific development progress beyond man's origin. And the next thing you know, there will be a legal battle staged within the corners of this state, that challenges even permitting anyone to believe that Jesus Christ was divinely born— that Jesus Christ was born of a virgin—challenge that, and the next step will be a battle staged denying the right to teach that there was a resurrection, until finally that precious book and its glorious teachings upon which this civilization has been built will be taken from us.

On Friday, the sixth day of the trial, Judge Raulston read his long ruling explaining why he intended to exclude the scientists from giving testimony before the jury. This caused Darrow to make some extremely critical and sarcastic remarks to the court, which could be considered in contempt. The judge was offended, and Darrow was required to post a bond for $5,000 while the judge decided what to do about Darrow's remarks. The trial day was short; the court adjourned at 10:30 a.m. The weekend must have been tense for all concerned. On Monday Darrow offered an apology and Judge Raulston responded: "My friends, and Col. Darrow, the Man that I believe came into the world to save man from sin, the Man that died on the cross that man might be redeemed, taught that it was godly to forgive and were it not for the forgiving nature of Himself I would fear for man. The Savior died on the cross pleading with God for the men who crucified Him. I believe in that Christ. I believe in these principles. I accept Col. Darrow's apology".

Among other items of business on Monday, a letter from the governor of Tennessee was read. It stated in part: "After careful examination I can find nothing of consequence in the books now being taught in our schools with which this bill will interfere in the slightest manner. Therefore, it will not put our teachers in any jeopardy. Probably the law will never be applied". The governor seemed to be making a strong point for the defense, but Judge Raulston ignored the letter and ruled that the courts, not the executive branch of government, would decided what the laws meant. "His opinion of what the law means, whether or not it would be enforced, is of no consequence at all in the court, and could not have any bearing, and I exclude the statement". The defense was not permitted to refer to the state's newly adopted biology textbook, which presented evolution as an extraordinarily important concept in biology. This was the day that the statements for the defense prepared by the scientists and religious leaders were read into the record.

Later in the day, one of the most astonishing events in the trial occurred. After a lengthy discussion of the many variants of the Bible, the defense wished to question witnesses on the *Bible*, and Bryan agreed to take the stand as an expert. The jury missed all of this. Darrow asked the questions and Bryan answered. Here is a sampling:

darrow: Should everything in the Bible be literally interpreted?

bryan: Yes, everything literally except some illustratively such as man being the salt of the earth.

darrow: Was Jonah literally swallowed by a whale?

bryan: I think it was a big fish.

darrow: Did Joshua make the sun stand still?

bryan: I believe what the Bible says.

darrow: Did the Flood of Noah occur and, if so, when?

bryan: Yes.

darrow: Do you know that the date of 4004 b.c. was determined from the generations given in the Bible?

bryan: I am not sure.

darrow: What do you think about the date?

bryan: I do not think about things I don't think about.

darrow: Do you think about things you do think about?

bryan: Sometimes.

darrow: Do you think the earth was made in six days?

bryan: Not six days of twenty-four hours. [This answer was greeted with gasps from the fundamentalists in the courtroom—they were literalists and if the Bible said days, that meant a day consisting of 24 hours.]

darrow: Was the first woman Eve?

bryan: Yes.

darrow: Was she made from Adam's rib?

bryan: Yes.

darrow: Where did Cain get his wife?

bryan: I do not know.

Each question was explored in detail, and slowly Bryan began to crumble. He proved to be a poor witness when it came to questions about the Bible, and he knew he was letting down those in the courtroom who believed in the inerrancy of biblical statements. The attorney general interrupted to ask the purpose of Darrow's line of questioning—hoping to put an end to it:

bryan: The purpose is to cast ridicule on everybody who believes in the Bible, and I am perfectly willing that the world shall know that these gentlemen have no other purpose than ridiculing every Christian who believes in the Bible.

darrow: We have the purpose of preventing bigots and ignoramuses from controlling the education of the United States and you know it, and that is all.

bryan: I am simply trying to protect the word of God against the greatest atheist or agnostic in the United States (prolonged applause). I want the papers to know I am not afraid to get on the stand in front of him and let him do his worst. I want the world to know (prolonged applause).

On Tuesday, July 21, 1925, the eighth and last day Darrow reminded the court that Scopes had taught evolution as charged, including that humans had evolved from a lower order of animals, so the jury should be instructed to find Scopes guilty. The jury was brought back into the courtroom, and Judge Raulston gave his instructions. The jury retired, deliberated for nine minutes, and found Scopes guilty. The judge set the fine at $100—an amount that was to prove critical as events unfolded. The court asked Scopes if he had anything to say. He did. "*Your Honor*. I feel that I have been convicted of violating an unjust statute. I will continue in the future, as I have in the past, to oppose this law in any way I can. Any other action would be in violation of my ideal of academic freedom—that is, to teach the truth as guaranteed in our constitution, of personal and religious freedom. I think the fine is unjust". So the case came to its close. Bryan had this to say:

Causes stir the world. . . . Here has been fought out a little cause of little consequence as a case, but the world is interested because it raises an issue, and that issue will some day be settled right, whether it is settled on our side or the other side. It is going to be settled right. There can be no settlement of a great cause without discussion, and people will not discuss a cause until their attention is drawn to it, and the value of this trial is not in any incident of the trial. . . . [T]his case will stimulate investigation and investigation will bring out information, and the facts will be known, and upon the facts, as ascertained, the decision will be rendered. . . . [N]o matter what our views may be, we ought not only to desire, but pray, that that which is right will prevail, whether it be our way or somebody else's.

Darrow was not so philosophical: "I think this case will be remembered because it is the first case of this sort since we stopped trying people in America for witchcraft because here we have done our best to turn back the tide that has sought to force itself upon this—upon this modern world, of testing every fact in science by a religious dictum. That is all I care to say".

Darrow had argued for what was right. Bryan had argued for what was righteous. Darrow had lost. He had hoped to show that the Butler Act was unconstitutional because it violated freedom of speech and religion and to call distinguished scientists and theologians for their testimony of what the theory of evolution and the Bible said about creation. Judge Raulston overruled all of these requests. It was admitted that Scopes had taught evolution, as charged; so unless the Butler Act could be shown to be unconstitutional, there could be no other verdict than guilty.

All in all, the Scopes trial was a bizarre affair. The textbook that contained the discussion of evolution had been selected years before by the State of Tennessee for use in its schools. Apparently this choice had never upset parents, whether they were fundamentalists or not. The stimulus for the trial was not local discontent but the desire of the ACLU in New York for a test case. A citizen of Dayton had to convince Scopes to say that he had taught evolution and so to be tried; Scopes was never asked what had really happened. Had he been asked, there would have been no case, because after the trial ended he confessed to a newspaper reporter that he had not been in school on the day evolution was discussed in his biology class. Another teacher had substituted for him!

Bryan had prepared a long speech to give before the jury, but he never had the opportunity. After Darrow finished his devastating questioning of Bryan about the *Bible*, the attorney general thought Darrow might savage Bryan once again and so decided it would not be wise to allow Bryan to give his speech even to redeem himself. Byran did give copies of the undelivered speech to the press after the trial ended. He spent the next few days traveling and making speeches. On Sunday, July 26, five days after the trial ended, Bryan was back in Dayton. After a hearty dinner, he died in his sleep. Some say this was in part a consequence of the deep humiliation he had suffered at the trial. Others thought his death was a consequence of overeating. Maybe both are true.

In a final attempt to have the Butler Act declared unconstitutional, the verdict in the Scopes trial was appealed to the Supreme Court of Tennessee, which was embarrassed by the whole affair and had a difficult time deciding what to do. Happily for the justices, they found a way to avoid a decision. The jury in the Scopes trial had returned a verdict of guilty, but they did not specify the fine. The amount was left to Judge Raulston, and he set it at $100. That was a big mistake. The constitution of Tennessee required that fines of more than $50 be determined by the jury. Therefore the Supreme Court declared a mistrial and reversed the decision that Scopes was guilty. The usual next step would have been to hold a new trial. However, the Supreme Court felt that nothing was to be gained by prolonging the life of this "*bizarre case*" and suggested that the attorney general forget the entire matter—which he did.

After Dayton

The Scopes trial did nothing to resolve the debate between the creationists and the evolutionists. The lawyers for the defense were unable to have the act declared unconstitutional either at the trial or upon appeal to the Supreme Court of Tennessee. The mistrial meant that the prosecuting lawyers had failed as well, because Scopes was not convicted. The Butler Act remained part of the law of Tennessee, but it was rarely invoked and was finally repealed in 1967. Other states, mainly in the southeast, considered bills banning the teaching of evolution. In most cases they did not pass. In Kentucky, for example, an antievolution bill was introduced, but it was laughed away—another bill introduced at the same time demanded that water should run uphill.

Some church leaders and church groups even came out in support of teaching evolution.

On balance, however, the Scopes trial proved to be a plus for the creationist movement. Bryan, as a chief spokesman and revered leader of the nation's creationists, gave national stature and brilliant oratory to the cause. Not unexpectedly, his followers were shocked and dismayed when under Darrow's searing cross-examination Bryan proved to be inept, especially in suggesting that the "*days*" of creation might not have been of 24-hour duration but possibly were very much longer. Bryan overcame this lapse from orthodoxy by conveniently dying less than a week after the trial ended. Darrow was blamed for contributing to this outcome, and Bryan became a martyr—a powerful plus for any religious cause.

The ascendancy of creationism had a chilling effect in the classroom. The solution for many teachers was to omit evolution altogether. Teaching in public schools is difficult enough, and it is not made easier when parents complain that their children's sacred beliefs are being undermined and that they are "*turning from God.*" Most teachers in the 1920s were not well prepared to teach evolution anyway—a situation that persists to this day. Textbooks reduced their coverage of evolution and placed any discussion of it at the very end of the book. This placement allowed a teacher to say that "we never had time to get to that topic." Rarely was evolution presented as the grand organizing theory that makes so many biological things and processes understandable.

A notable example of the effect of the Scopes trial can be seen in the history of the textbook Biology for Beginners by Truman J. Moon, which was the predominant high school biology textbook in the nation for decades. In the 1921 edition, published four years before the Scopes trial, the frontispiece carried a full-page portrait of Charles Darwin, and a short chapter titled "The Development of Man" began as follows:

With an egotism which is entirely unwarranted, we are accustomed to speak of "man and animals" whereas we ought to say "man and other animals," for certainly man is an animal just as truly as the beasts of the field. . . .

As soon as man became intelligent enough to make comparisons between himself and other animals, the resemblances became apparent and led to the idea that some relationship must exist with lower forms. Two thousand years ago the Greeks discussed this fact and advanced various theories to account for it.

Very gradually, information accumulated, and the idea of relationships developed into the theory that not only man but all other living things, both plant and animal, are not only related, but actually descended from common ancestors. This is called the theory of *descent*, or *evolution*.

Moon then listed the evidence for evolution—such as rudimentary organs, embryological resemblances, *homologous organs*, geological data, domestication of animals and plants—and followed this with a family tree showing the evolutionary relations of human beings and the great apes. There was no discussion of the forces that Darwin thought were responsible for evolution—genetic variation and natural selection. After the Scopes trial, the picture of Darwin was removed from the front of the 1928 edition of Moon's textbook, and the 1933 edition omitted evolution entirely.

The biology actually taught from the late 1920s to the 1960s consisted mainly of detailed descriptions of the structure of organisms "from amoeba to man." That approach not only was far easier for the teacher to implement but also avoided unpleasant confrontations with parents about evolution. Although professional biologists at that time might have desired an adequate treatment of evolution in textbooks, decisions about content were made primarily by publishers, whose major concern was meeting the needs of the largest sector of their market. Consequently they conveniently avoided controversial subjects.

A movement in the late 1950s throughout the United States to improve science education, especially in high schools, was stimulated by Russia's launching of the satellite Sputnik in 1957. American leaders saw that accomplishment as indicative of better science education in the Soviet Union. Separate national committees for physics, mathematics, chemistry, and biology were formed to study the problem of improving science instruction, and each included university scientists, high school teachers, and school administrators. The biologists formed the Biological Sciences Curriculum Study (BSCS) and, with support and encouragement from the National Science Foundation, prepared experimental textbooks intended for use in tenth-grade biology courses.

The BSCS Steering Committee proposed to get away from the "parade-of-the-animal-and-plant-kingdoms" approach, to stress concepts and experimental science, and to encourage the personal involvement of students in their learning—especially in the laboratory. The two areas in biology that had been prominently ignored in high school biology courses were human reproduction and evolutionary biology.

Both were to be adequately treated in the new BSCS books. One of the members of the BSCS Steering Committee, the distinguished geneticist and Nobel laureate H. J. Muller, proclaimed that "a hundred years without Darwin are enough."

The BSCS produced three textbooks, each emphasizing a different approach to biology but all including evolution and human reproduction. These controversial topics could be included because the contracts made with the publishers contained a clause giving full control of content to the BSCS. Some publishers were not happy with this arrangement, yet the extensive publicity surrounding BSCS and the other national curriculum projects suggested large sales for the new kind of books. Evolution became a major theme in BSCS biology books in the 1960s and, having been given the imprimatur of what was seen as a national reform effort, it spread to other biology textbooks as well. For a few years it seemed as though the "hundred years without Darwin" were over. Not quite.

10

GENES AND INTERPRETATION

You are working in a building with lots of rooms. There is something slightly unusual about the rooms because each has a door with several locks on it, designed in such a way that to open a door you need a key that fits at least one of its locks. It so happens that you mainly use three of the rooms and have a key that matches a lock on each of their doors. But you now need to enter a new room, room 23, which your key doesn't fit. Frustrated, you try and change your key, filing off some bits here and there. Eventually you manage to modify your key so that it now fits one of the locks on room 23, allowing you to enter. The only trouble is that you are now locked out of the other three rooms because your key no longer fits them. In adapting the key to fit a different type of lock, you have sacrificed its ability to work on the others.

There is a much simpler way of solving the problem which avoids all this: don't mess around with the key but get the locks on room 23 changed. You ask the locksmith to install an additional lock on room 23 that matches your key. By installing the new lock, the ability of your key to work on the other doors is not affected, so you are now able to open the new room as well as the original three. Similarly, if it happens that a door currently opened by your key, say on room 12, needs to be closed to access by you (perhaps for security reasons), it would be a mistake to change your key because then it would no longer work on any of the doors you still need to enter. It would be better to simply remove the lock on room 12 that your key fits. In other words, if you want to modify a system involving components that recognize or match each other, like locks and keys, it can make a big difference which way round you change the components. In this case, it is much easier to change the locks than the keys.

If we imagine the locks being continually modified in this way over a period of time, with some locks being added or taken away from certain doors, we will end up with a building that has a particular pattern of locks on its rooms which can be opened with a set of matching keys. A particular key may open a set of doors, say on rooms 5, 10 and 23. Now without knowing the history of the building, it may seem that the key has been designed to open just these doors.

After all, you need the key to open or close a door, so it seems to have more control than the lock. But when we know the building's past, it is clear that the keys have only played a rather passive role in the design. It is the locks that have been changed and modified, not the keys. The keys have in a sense been at the mercy of where the locks they fit have been placed, rather than the other way round. We might almost say that by acquiring a matching lock, certain doors have allowed themselves to be opened by a key, rather than the key being able to dictate which doors it can open.

Evolution of Interpretations

In the previous chapter, we saw how a *master protein* (hidden colour) can bind to a particular site in the regulatory region of an interpreting gene, switching the gene on or off, much as a key fits into a lock. Now, just as it was easier to change the locks than the keys in our building full of rooms, much of biological evolution has involved changes in the binding sites within *regulatory regions* (*locks*) rather than in the *master proteins* themselves (*keys*). Because a typical master protein might bind to as many as one hundred different interpreting genes, an enormous constraint is imposed on the extent to which the shape of this master protein can be modified during evolution: any significant change may jeopardise the expression of all one hundred genes it normally binds to, most likely with disastrous consequences for the development and survival of the organism. In contrast, by changing a binding site in the regulatory region of a gene, only the expression of that gene will be directly affected. For this reason, evolutionary changes are often likely to involve mutations in the sites within regulatory regions, rather than alterations in the regions coding for the master proteins themselves.

An example of a change in a regulatory region might be the creation of a new binding site. Binding sites are quite short stretches of DNA, typically six to ten bases long. In a regulatory region a few thousand bases long, it is not too improbable that a chance mutation altering one or two bases in the DNA could create a new binding site for a master protein, or at least something that came reasonably close to a new binding site. This sort of mutation might start to couple the interpreting gene to a different master protein, modifying the gene's pattern of expression. If this new pattern proved advantageous for the organism, further mutations in the regulatory region might then be selected for to improve the match, creating an even better binding site. A new binding site (*lock*) has evolved to match a master protein (*key*) that was already around. It is also easy to see how a mutation could lead to the loss of a binding site in a regulatory region. A change in just one base in the DNA sequence of a binding site could mean that the master protein that normally recognizes it can no longer bind. The interpreting gene would then no longer respond to this particular master protein.

Thus, the combination of binding sites in the regulatory region of a gene is something that has gradually evolved. During the course of evolution, particular binding sites have arisen or been lost, changing the way interpreting genes respond to the patchwork of master proteins.

Master proteins are only masters in the sense that they can influence the activity of many genes, just as a key might open many doors. They are not masters in the sense of dictators, having evolved all the information that decides which interpreting gene should be on or off. This is because of the way the system has evolved, through genes modifying their response to the master proteins, rather than the master proteins evolving more and more complex shapes that allow them to dictate to more and more genes.

At the early stages in the evolution of a master protein, when it may bind to just a few genes, there may be quite a bit of room for change. But as more interpreting genes evolve suitable binding sites and come under the influence of the master protein, the possibilities for change become more limited. It then becomes increasingly more likely that altered patterns of gene expression involve changes in interpretation, rather than changes in the master proteins themselves.

We can now see that genes interpret hidden colours, in the sense: (1) The hidden colours provide a frame of reference, a distribution of master proteins of various types. (2) Each gene responds to this pattern selectively, being expressed at various times and places in the organism according to the set of binding sites in its regulatory region. A different combination of binding sites leads to a different pattern of expression.

(3) The particular selection made in each case is historically informed, depending on a series of historical events that have led to one set of binding sites in the regulatory region rather than another.

Families of Colour

Although the evolution of *master proteins* is constrained, there is one very important way in which these restrictions can be partially overcome: through a process called *gene duplication*. This occurs when a mistake is made during the copying of DNA. Remember that DNA is normally copied once every time a cell divides. Occasionally an error is made in the copying process such that one stretch of DNA ends up being copied twice instead of once. The details of how this occurs need not concern us here: what matters is that it sometimes results in an extra copy of a gene being incorporated in the DNA. This means that if we have an organism with one gene for a master protein, very occasionally a descendant will be produced with an extra copy of the gene. The descendant now has two copies of the gene for a hidden colour. These two copies will have exactly the same sequence of bases in their DNA: they will be 100% identical.

Duplication seems to be of little consequence at first, but in the longer term it can provide greater evolutionary flexibility as the duplicate copies diverge. This is for the same reason that if you have two copies of a key, you can tinker around with one of them without jeopardizing your ability to open doors, because the other copy acts as a backup. If you have two copies of a gene, some mutations that might normally be detrimental to the organism could be allowed

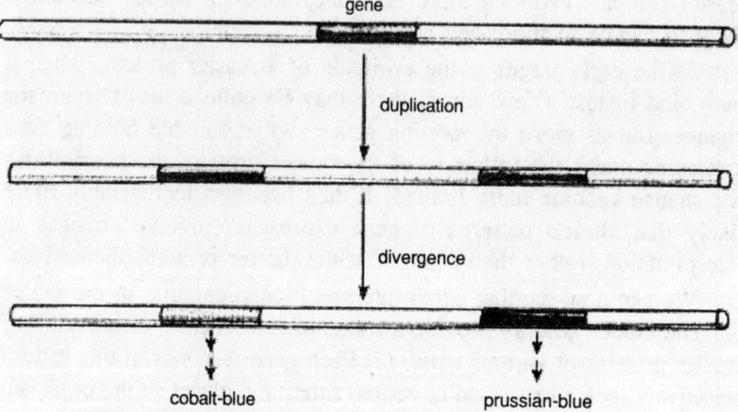

Fig. 10.1. Duplication and divergence of a gene coding for a master protein, resulting in genes for two different types of blue hidden colour.

because the genes act as backups for each other. Mutations would be expected to accumulate over a period of evolutionary time, so that the duplicate genes eventually start to diverge in sequence. The duplication has been followed by divergence between the copies, so the DNA sequences of the duplicate genes are no longer 100% but, say, only 90% identical (that is, out of every ten bases of DNA sequence, there is now on average one difference, just as the word convection only differs in one out of ten letters from conviction). As the DNA sequences diverge, so the proteins encoded by each of the duplicates may also start to diverge. Perhaps this would lead to their shapes becoming slightly different in some way, so that they now bind to regulatory regions with a slightly different specificity. Having started off with one hidden colour, two different versions have evolved. The two hidden colours will be closely related to each other Many of the hidden colours affecting identity in flowers and flies are thought to have arisen by duplication and divergence. This became very dear when the identity genes needed for the hidden colours were isolated in the early 1980s. Once a gene has been isolated, the sequence of its DNA and encoded protein can be determined. By comparing the sequences of different identity genes, it soon became apparent that they had arisen by gene duplications. For example, the eight fly genes needed for colours a all have similar DNA sequences.-h The percentage similarity between the eight genes is particularly high in one stretch of their DNA, about 180 bases long, named the homeobox. The homeobox provides a sort of common signature, showing that the eight genes all started as duplicates of each other. Because of this, the identity genes of the fly are also sometimes referred to as homeobox genes, as they all share this region of similarity in their DNA. This does not mean that the homeobox region is identical in the different genes; they each have slight differences in this region due to divergence. It is just that the homeobox is the region of greatest similarity between the genes.

The reason that the homeobox is thought to be so well conserved between duplicates is that it codes for the part of the master protein (called the homeodomain) that makes direct contact with the DNA: the region of the master protein that fits into the binding site (equivalent to the part of a key you insert into a lock). Any major alterations in the sequence of this region are likely to disrupt the ability of the master protein to work at all, and alterations have therefore been selected against during evolution.

Because the *master proteins* encoded by these genes are similar or related to each other, we can symbolize them as a family of related

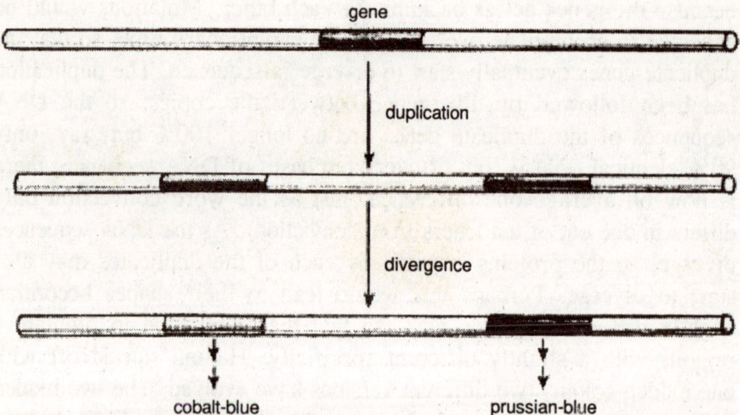

Fig. 10.2. Cluster of identity (homeobox) genes in fruit fly DNA with the corresponding hidden colours they code for.

hidden colours, say various types of green. We can replace the colours a with types of green that begin with the same letter: colour a becomes-h apple-green, colour *b* is now bottle-green, *c* is cyprus-green, *d* is deep-green, *e* is emerald-green, *f* is forest-green, *g* is grass-green and *h* is herb-green. The fruit fly is divided up into territories coloured with various types of green, starting with apple-green at the head end and finishing with herb-green at the tail *e* Each type of green territory corresponds to a region of cells where and particular type of master protein is made.

The identity genes needed for the various green colours are arranged in two clusters in the DNA: five genes are in one cluster and three in the other. You may notice that the order of the genes in the clusters is the same as the order of hidden colours from head to tail in the organism. For example, looking at the cluster on the left, the gene for apple-green is followed by the gene for bottle-green, followed by the gene for cyprus-green and so on, paralleling the order of the corresponding territories of hidden colour from head to tail in the animal. It is still not clear why the order of these genes in the DNA should correspond so nicely with their order of expression in the organism, but it most likely has to do with the way the duplications have evolved. Many of the identity genes affecting whorls of flower organs have also arisen by duplication and divergence. These genes each contain a similar stretch or signature in their DNA sequence. This signature is not the same as the one in the fly identity genes, and is therefore given a different name: the *MADS-box*.

Instead of greens, we could represent the set of master proteins encoded by these genes as reds. The hidden colours of the flower previously referred to as a, b and c can now be replaced by amarone-red, burgundy-red and claret-red respectively (fortunately the wine trade has provided us with many names for reds). The flower -bud can be thought of as containing concentric territories of red colour: starting with amarone-red in the outermost whorl (sepals), then amarone-red + burgundy-red (petals), then burgundy-red + claret-red (stamens), and ending with claret-red in the centre (carpels). This pattern of hidden colours is what gives a separate identity to the various whorls of flower organs.

Let me summarize the main points so far. Flies and flowers contain a set of identity genes that are expressed in various regions of the organism to produce *master proteins*. This distribution of master proteins is equivalent to a map or patchwork of hidden colours. Many of the master proteins are related to each other because the various identity genes arose by duplication and divergence, and this can be symbolized by the red (flowers) or green (flies) families of hidden colour. The map of hidden colours provides a frame of reference that can be interpreted by many genes through their regulatory regions. The combination of binding sites in a regulatory region acts like a specific molecular antenna, responding to the pattern of hidden colours in such a way that each of these genes comes to be expressed at certain times and places in the organism.

Genes and Language

It is useful to compare this view of *genes* with the way our own language works. Each gene, made up of a sequence of DNA bases, is often compared to a word comprising a series of letters. The equivalent of all the thousands of genes in the total DNA of an organism might then be a large dictionary with a vocabulary of thousands of words. There is, however, a fundamental difference between dictionaries and DNA when it comes to expressing their contents. To express a word from a dictionary, someone needs to look up the word and pronounce it. The word itself does not carry information that tells you whether to say it or not—this comes from the reader who is using the dictionary. A *gene*, however, does carry information in its regulatory region that determines when and where it is expressed.

The gene contains a molecular antenna, a series of binding sites, ensuring that it is expressed in some cells and not others. It would be as if each word had a large prefix that ensured it was pronounced at certain times and places. Although analogies with the written word break down here, there is in my view a better type of linguistic comparison: with the way we use words in our head

When you talk, or experience a train of thought, the words seem to come automatically. Suppose you have a thought like 'I wonder how bees look at flowers: You do not look up each word like 'I', then 'wonder', then 'how', in a mental dictionary, because to do so you would first have to know what words you wanted to look up: to look up 'wonder' in your head you would already have to know that 'wonder' is the word you wanted, defeating the whole point of the exercise. The words we use are in a sense stored in our brain, but they occur to us under particular conditions rather than being something we look up in a mental reference library. We are most conscious of this when a word is on the tip of our tongue and we have difficulty in recalling it. We have to wait until the word comes to us almost of its own volition. Thoughts are not something we plan and then execute by looking up the appropriate words, we just have them. We can of course plan to think about something, like 'I am going to spend the next hour thinking about bees'; but we do not plan the thoughts we will then have about bees and retrieve the words accordingly, because to do so would mean that we had already had the thoughts. Rather, we might start by contemplating some aspect of bees, and this would lead to other thoughts and words coming to mind. We experience a wandering train of thought rather than a planned series of events.

By analogy with *genes*, we might notionally divide each word in our brain into two parts. One part has to do with what gets expressed as the word occurs to us, and is responsible for how the word 'sounds' in our head. By '*sound*' here, the experience of having the word in our conscious mind, irrespective of whether we say it aloud or not. This would be equivalent to the coding region of a gene producing a particular protein. The second part of a word would determine when the word occurs to us, ensuring that each word comes to our mind under certain conditions. This would correspond to the regulatory region of a gene. It is as if each mental word carries information that leads to its being expressed or manifesting itself in our consciousness according to the conditions in our brain, rather than just being an entity that we retrieve. Words in our mind are not the same as those written down on a page; they are networked or locked into the thinking process. Of course the way they are locked in is not immutable: it can change as our experiences and mental processes develop. At any one time, the '*regulatory part*' of each word is historically informed, depending on our previous learning experiences.

Mental words are as simple as a linear sequence of subunits in a gene. We do not yet know how words work in our mind, but they most likely reflect a complex set of interactions between cells in our brain. It may be that these interactions would defy being simply broken down into the equivalent of regulatory and coding parts of a gene. My reason for drawing this comparison between mental words and genes is not to give an oversimplistic view of the mind, but to give us a better sense of how genes work than is implied by the notion of a dictionary. There is no independent reader dipping into the gene volumes held within each cell. *Genes* carry information that leads to their being expressed at certain times and places.

Now of course the whole process of thinking is remarkably interactive. Every word or thought that occurs to you leads to new words coming to mind. There is an ever changing state of mind in which each word or thought feeds off the previous ones. This is also true for genes. Genes come to be expressed by interpreting hidden colours; and these hidden colours or master proteins themselves depend on a set of genes (*identity genes*). Genes feed off each other much as words do. But how the genes coding for the hidden colours themselves get to be expressed in a pattern. It is all very well saying that genes interpret a complex patchwork, but what sets up the patchwork to begin with? Assuming that the pattern of hidden colours is already given. As we shall see, the production of this pattern depends on further interactions between genes and proteins.

11

DOUBTFUL IDENTITY

At three in the morning on the third of September, 1786, Johann Wolfgang Goethe jumped into a coach, assumed a false name, and set off for Italy. Goethe had just turned 37. In his youth, he had achieved great success with the publication of a tragic novel, The Sorrows of Young Werther. The book was so popular that a cult industry rapidly grew around it. There were Werther plays, operas and songs; even pieces of porcelain were made showing Werther scenes. In spite of his outstanding literary success, Goethe chose at the age of 26 to serve for a period in the court of Weimar, at the invitation of the Duke. At various times during the next eleven years he assumed responsibilities for the mines, the War Department, and the Finances of the Duchy. However, life in Weimar eventually proved too restrictive and by the time he was 37 Goethe felt impelled to escape incognito to a new environment.

Goethe travelled around Italy for about twenty months. During this time he developed various scientific theories concerning the weather, geology and botany. It may come as a surprise that so famous a poet should have concerned himself with science. Goethe, though, had far-ranging interests in nature. His scientific work was particularly important to him, and he dedicated much of his time to it. The aspect that most concerns us here is an important botanical idea he had during his Italian journey.

A UNIFYING THEME

To understand Goethe's idea and how he came to it we need to go back a few years to a discovery he made during his period at Weimar at the age of 34. Goethe had been struck by fundamental

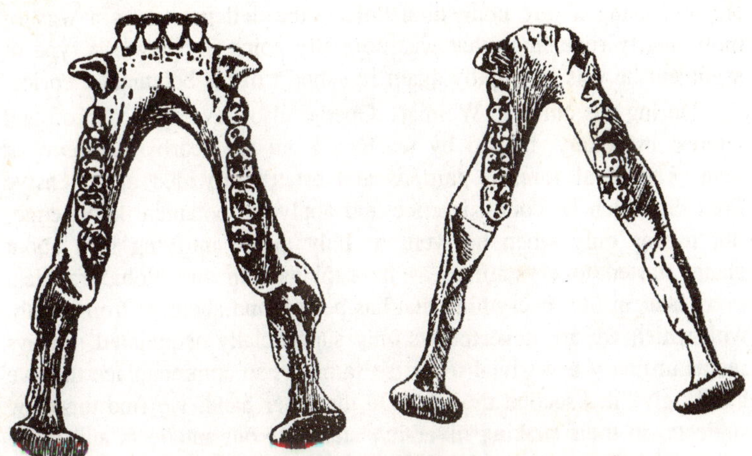

Fig. 11.1. Lower jaws of chimpanzee (left) and of Dryopithecus africanus.

similarities in the structures of different organisms and became convinced that they were all formed in a common way. One of the most obvious illustrations of this was the similar arrangements of bones in the skeletons of many different animals. For instance, the human thigh bone, or *femur*, had an easily identified counterpart in a dog, bull, lion or any other mammal. However, although such a one-to-one correspondence could be established for most bones in the body, there were some apparent exceptions. For example, monkeys had a bone in the middle of their face, called the *intermaxilla*, which appeared to be lacking in humans (this bone is also known as the *premaxillary*). This was often taken to be an important distinguishing mark that separated man from ape. But Goethe's belief in a fundamental unity between organisms encouraged him to look much more closely at the human skull. Eventually he discovered that the intermaxillary bone was also present in man but it had been overlooked because it was tucked away in the upper jaw and was closely joined with other bones. His conviction in the commonality of forms had led him to discover something that others had missed. He was able to show that rather than being a distinguishing mark, the intermaxillary bone was actually a connecting link that unified man with other animals.

One piece of evidence that Goethe used to support his identification of the bone came from abnormalities. He noted that in individuals born with a cleft palate, the cleft almost always ran along the join between the proposed inter-maxillary region and the surrounding bones, pointing to the intermaxillary bone as being a separate entity.

He was using a rare congenital defect, the deft palate, as a way of more dearly revealing what was normally going on. It was a type of argument he was to employ again in support of his botanical theories.

During his time at Weimar, Goethe also developed a profound interest in botany, helped by teachers from the nearby Academy at Jena. The local forests, gardens and estates provided an extensive flora on which he could practice and apply his botanical knowledge. But it was only when he went to Italy that a unifying idea about plants started to crystallize, as he explained in an autobiographical essay later in life: everything that has been round about us from youth, with which we are nevertheless only superficially acquainted, always seems ordinary and trivial to us, so familiar, so commonplace that we hardly give it a second thought. On the other hand, we find that new subjects, in their striking diversity, stimulate our intellects and make us realize that we are capable of pure enthusiasm; they point to something higher, something which we might be privileged to attain. This is the real advantage of travel and each individual benefits in proportion to his nature and way of doing things. The well-known becomes new, and, linked with new phenomena, it stimulates attention, reflection and judgement.

Exposed to a new flora during his Italian journey, Goethe was stimulated to think more deeply about plants. As with his work on skulls, he was searching for a fundamental unity that lay behind the surface of things. He came to realize that there was a single underlying theme to plants, epitomized by the leaf. It seemed to him that the same theme occurred again and again throughout the life of every plant: While walking in the Public Gardens of Palermo, it came to me in a flash that in the organ of the plant which we are accustomed to call the leaf lies the true *Proteus* (Proteus is a sea god of Greek and Roman mythology fabled to assume various shapes.) who can hide or reveal himself in all vegetal forms. From first to last, the plant is nothing but leaf, which is so inseparable from the future germ that one cannot think of one without the other. Anyone who has had the experience of being confronted by an idea, pregnant with possibilities, whether he thought of it for himself or caught it from others, will know that it creates a tumult and enthusiasm in the mind, which makes one intuitively anticipate its further developments and the conclusions towards which it points.

On returning to Germany, Goethe wrote up his idea in an essay, The *Metamorphosis of Plants*, published in 1790. He began by describing the typical life of a plant. After germination of the seed, a tiny shoot bearing one or two small leaves emerges from the ground. As the seedling grows, foliage leaves are successively produced, spaced out around the axis of the stem. At this stage all there is to the plant is stem and leaves (Goethe was not concerned with roots in his account). Eventually, however, the plant starts to form flowers. The question was how flowers might be related to the rest of the plant. Goethe proposed that the different parts of a flower were fundamentally equivalent to foliage leaves; it was just that instead of being spaced out along a stem, the parts of a flower were all clustered together.

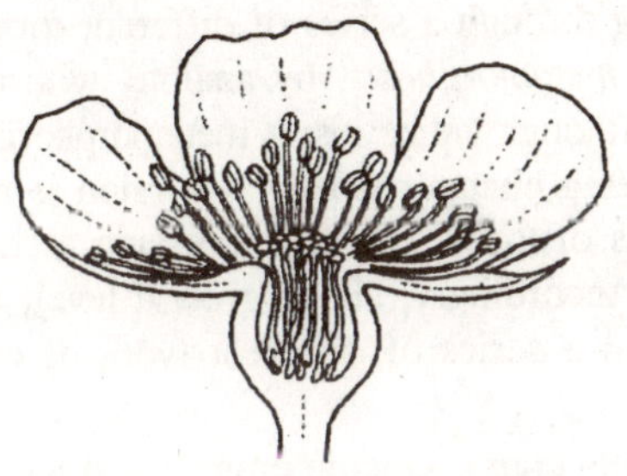

Fig. 11.2. Rossa: flower cut lengthwise.

Take a flower and look at its component parts. You will find that there are several types of organs, clustered around each other in concentric rings or whorls. By whorl, a region or zone of the flower that normally includes organs of one type (this is not quite the same as a botanist's definition but it will be more useful for our purposes). Many flowers have four whorls of organs. The outermost whorl comprises the sepals, usually small green leaf-like structures that protect

Fig. 11.3. Floral diagram.

the flower when it is in bud. Within these is a whorl of petals, usually the most obvious and attractive parts of a flower. Next come the stamens, the male sex organs that bear pollen. Finally, in the centre are the carpels, the female organs that when pollinated will grow to form fruits containing seeds.

Goethe proposed that the floral organs, as well as all the foliage leaves, were simply different manifestations of a common underlying theme. This theme could be realized in different ways during plant growth, first as foliage leaves, then as the organs of a flower: sepals, petals, stamens and carpels. It seemed as though an underlying organ was simply passing through a series of different forms. He called this process of change *metamorphosis*, by analogy with the changes many insects experience (though unlike insect metamorphosis where the whole organism undergoes a change, Goethe's version is more abstract and refers only to parts of the organism expressing a change, the various leaf-like organs). Accordingly, above ground level, a plant was made solely of stems and a series of different types of organs based on a common theme.

In support of his claim, Goethe emphasized the many similarities between flower organs and foliage leaves. It is perhaps not too difficult to imagine that sepals are equivalent to leaves because they usually have a very leaf-like appearance. Petals are also not so different from leaves, give or take a bit of shape and colour. But what about the sex organs? The male organs (stamens) do not bear any obvious resemblance to leaves. In the case of the female organs (carpels) we sometimes get a faint leaf-like appearance when they have been fertilized and grow into fruits or pods containing seeds: a pea pod could be thought of as a leaf that has been folded lengthways and had the edges stuck together. But what about a tomato? Slice a tomato cross-wise and you will see two or more segments, each containing seeds. Is a tomato several leaf-like organs joined together? The tomato segments do not look like leaves, so it is not at all obvious that they are the same sort of thing. As with his studies on the human skull, Goethe turned to abnormalities to help resolve the issue.

Helpful Monsters

Monstrous flowers are curiously attractive. For years gardeners have selected varieties with extra petals, sometimes called double-flowered forms. Roses, for example, have only five petals in the wild, yet many of the commonly cultivated garden varieties have many more than this. They have been selectively bred for their appeal to humans.

In some cases, these abnormal flowers have extra petals at the expense of sex organs, so they can no longer reproduce properly by sexual means (many of them are propagated vegetatively, by taking cuttings).

Although considered attractive to gardeners, most botanists viewed these abnormalities with suspicion, as unruly freaks of nature that would not repay further study. The eighteenth-century philosopher Jean Jacques Rousseau, also a keen botanist, warned young ladies against the dangers of such flowers: Whenever you find them double, do not meddle with them, they are disfigured; or, if you please, dressed after our fashion: nature will no longer be found among them; she refuses to reproduce any thing from monsters thus mutilated: for if the more brilliant parts of the flower, namely the corolla (*petals*), be multiplied, it is at the expense of the more essential parts (*sex organs*), which disappear under this addition of brilliancy.

Rather than shunning these monstrosities, Goethe realized that they could provide important clues to understanding how flowers normally form. To Goethe, the monstrous flowers with extra petals in their centre suggested that the sex organs could somehow be transformed into petals. Surely this showed that the different organs of a flower were interconvertible and so fundamentally equivalent. If this conclusion was granted, then the obvious similarity between foliage leaves and at least some of the flower organs (*sepals* and *petals*) indicated that all of the organs of a plant should be lumped into the same equivalence group. The various parts of a flower were equivalent to each other and to other types of leaves; they were all variations on a common theme. As further confirmation of this idea, Goethe cited abnormal roses which, instead of sex organs, had an entire shoot emerging from their centre, bearing petals and leaves. Here was a dear illustration of the equivalence between floral organs and leaves.

When Goethe wrote his essay on plant metamorphosis, he was not aware that some of the ideas had been arrived at twenty years before him, by Caspar Friedrich Wolff. Wolff was one of the founding fathers of the theory of epigenesis, the view that organisms develop by new formation rather than being preformed in the egg. At the age of 26, Wolff had produced a doctoral dissertation at the University of Halle, Theoria Generationis, which was remarkable in its scope and insights for having such a young author. It included a range of original microscopic studies on the development of plants and animals. From his plant work, he had been struck by how various parts, such as leaves and floral organs, arise in a similar way at the growing tips of the plant (Wolff was the first to describe the plant growing tip). A few years later, in 1768, he considered this in the light of abnormal flowers: one observes that the stamens in the *Linnaean Polyandria* [species with many stamens in their flowers] are frequently transformed into petals, thereby creating double flowers, and conversely that the petals are transformed into stamens; from this fact it may be concluded that the stamens, too, are essentially leaves. In a word, mature reflection reveals that the plant, the various parts of which appear so extraordinarily different from one another at first glance, is composed exclusively of leaves and stem, inasmuch as the root is part of the stem.

Wolff had come to the same conclusion as Goethe: the various parts of a flower could be thought of as equivalent to leaves, and thus the whole plant above ground was made up of only stem and leaf-like organs. Later on, Goethe came across this work and acknowledged Wolff's precedence. Nevertheless, Goethe developed the idea of the equivalence of plant organs much more extensively than Wolff, and put it forward more coherently as a theory of plant development.

The reception of Goethe's theory was mixed. Some biologists regarded his ideas as of the utmost importance, and viewed him as a founding father of morphology (Goethe coined the term), the scientific study of shape and form. Others were less generous and saw Goethe's contribution as over-idealistic, trying to make nature conform to his poetic views, rather than being a serious scientific theory based on hard facts: they were the dabblings of an amateur rather than an important scientific effort. Goethe's own view was that his work on science was much more than a mere adjunct to poetry. He took his scientific studies very seriously and continued with them for the rest of his life, dedicating much of his later time to the study of optics.

One of the problems with assessing Goethe's botanical ideas has been that, until quite recently, his theory could not be followed up experimentally. He was much more concerned with giving a general intuition of how plants were formed than with laying the foundations of an experimental programme of investigation. It was only with the advent of new approaches to the study of flower development that many of his ideas have come to be appreciated again from a fresh perspective.

Identity Mutants

Many of the flower abnormalities of the type described by Goethe are caused by *mutations* in particular genes. Their significance became much clearer during the 1980s, when systematic collections of such mutants were obtained by screening many thousands of plants for exceptional individuals with abnormal flowers. The screens were mainly carried out in two species: *Arabidopsis thaliana* and the snapdragon, *Antirrhinum majus*. To show how these studies helped illuminate the nature of floral monstrosities, Three important classes of mutant that emerged from these screens, called a, b and c have been described.

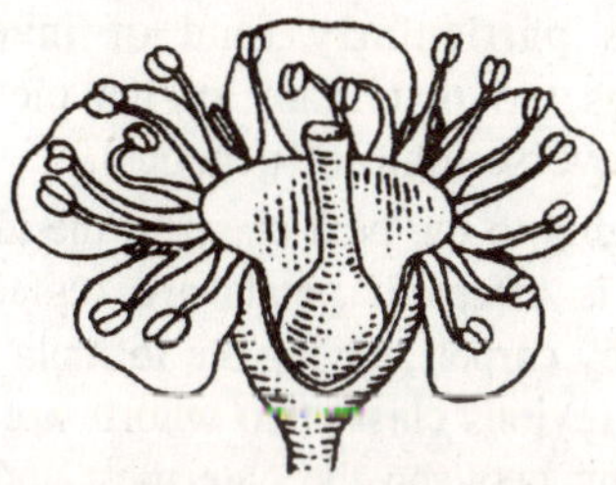

Fig. 11.4. Flower cut from one side.

Remember that a flower normally has four concentric whorls of organs, which proceed from outside to inside in the order *sepals*, *petals*, *stamens* and *carpels*. In mutants of class a, the sepals and petals, which normally occupy the outer two whorls, are replaced by sex organs: carpels grow in place of sepals, and stamens in place of petals. If we were to give a formula for the normal flower as sepal, petal, stamen, carpel, the class a mutant would be carpel stamen, stamen, carpel. In other words, structures that are normally restricted to the inner regions of the flower, the stamens and carpels, have now taken over the outer positions as well. It should be emphasized that this does not involve any organs actually moving or changing position. Rather, the outer organs develop with an altered identity, as carpels and stamens rather than sepals and petals. Each organ grows and develops in the same location as in a normal flower, but the organs in the outer whorls assume the same identity as those that are normally found in the inner whorls.

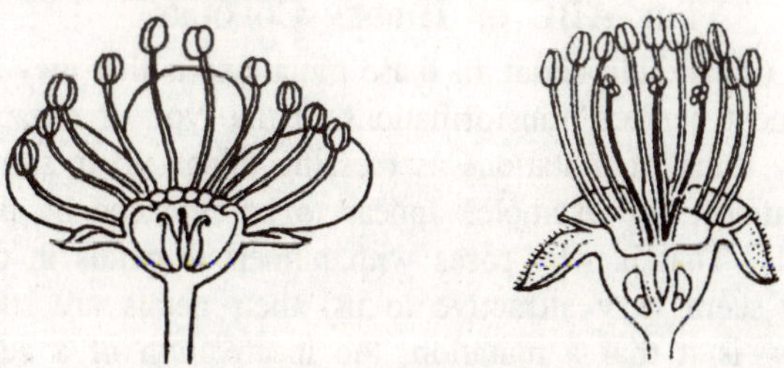

Fig. 11.5. Flower cut lengthwise.

The situation is somewhat reminiscent of a painting by Rene Magritte, showing a pair of shoes partially transformed into feet. Shoes are worn on the outside of our feet, yet here they acquire some of the features of what is normally within them, some human skin and toes. A structure on the outside has started to assume the identity of something that is normally within. In this case, however, the transformation is purposely left incomplete, so that only part of each shoe resembles a foot. (Magritte was particularly fond of inverting the normal arrangement of things, so that many of his pictures provide good illustrations for the reverse logic of genetics).

In mutants belonging to the next class, b, the identity of a different pair of organ types is affected: petals are replaced by sepals, and stamens are replaced by carpels, giving the formula sepal, sepal, carpel, carpel. As with the previous class, two whorls are affected but in this case it is the pair lying between the outermost and innermost whorls, where the petals and stamens normally form.

In mutants of class c, the two inner whorls of the flower are affected: stamens are replaced by petals, and carpels are replaced by sepals, giving sepal, petal, petal, sepal. This is essentially the opposite of class a mutants: inner reproductive organs are now replaced by outer sterile organs. Some garden varieties with extra petals may belong to this class. In some cases, you can get numerous petals in this way because the normal flower contains many stamens, each of which is replaced by a petal. (There are some additional complications with interpreting garden varieties. In some cases the transformations towards petals may not be complete, so you get only a proportion of the sex organs being replaced, sometimes imperfectly. This may be because the mutations have not fully inactivated the relevant gene. A further complication is that class c mutants can also have extra whorls within the flower, on top of the usual four, for reasons that are not yet fully understood.)

The ABC of Hidden Colours

What is remarkable about all these mutations is that they seem to result in almost perfect transformations in the type of organ made. We normally think of mutations as messing things up in some way, but here stamens, for example, appear to be replaced by perfectly formed petals. That is why roses with numerous petals in place of stamens can seem very attractive to us: their petals are still well-formed. How is it that a mutation, the inactivation of a gene, can lead to such a neat conversion?

We can get a helpful insight by considering a parallel situation in language. In many cases, if you remove a word from a sentence, the sentence will become grammatically incorrect and meaningless. As with many mutations, you end up with a mess. But there are some words that can always be removed without such ill-effects. Take the word please. Parents spend many hours indoctrinating children to say please. 'I want more juice' . . . 'What's the magic word, dear?' 'I want more juice please.' Both of these child's requests make perfect grammatical sense; it is just that one is considered rude and the other polite. The word please has a particular type of role: it provides a way of distinguishing between polite and rude sentences, rather than being essential for their grammatical structure. It is an arbitrary convention that requests are impolite unless they include the word please. We might say that without please, rudeness is assumed by default. The notion of a default allows us to see how the removal of a word can influence the significance of a sentence, whilst at the same time preserving its grammatical correctness. (It strikes me that our common convention is very inefficient: it would be better if the default state was polite and we should have to add extra words like you numskull to make a sentence rude, avoiding the needless waste of energy on teaching children to say please every time they ask for something.)

In a similar way, the genes affected in the mutant flowers have a special type of role that can be understood in terms of defaults. To see how this works, A simple model that was designed to account for the three mutant classes, a, b and c. The basic elements of this model were arrived at independently by two research groups in the late 1980s: Elliot Meyerowitz, John Bowman and colleagues working on Arabidopsis at Caltech, California; and Rosemary Carpenter and me working on Antirrhinum at the John Innes Institute, Norwich. There are various ways of presenting this model. It is important to bear in mind that these are abstract rather than real colours. Their only justification at this stage is to provide a convenient way of explaining the different types of floral mutant.

According to the model, the flower can be symbolized as four concentric rings of hidden colour, corresponding to the four whorls of organs: sepals, petals, stamens, carpels. These colours are themselves built up from a combination of three basic colours, called a, b and c. The outermost ring is coloured a, the next ring in is coloured with the combination a + b, third in is b + c, and finally c is in the centre.

These basic colours and their combinations therefore give a different colour signature to each whorl. Starting from the outer whorl and moving towards the centre, the combinations are: a, ab, bc, c, representing the identities sepal, petal, stamen, carpel respectively.

Now the key feature of the model is that if you remove one or more colours, the identity of the organs will change to a default determined by the remaining colours. Suppose, for example, that colour b is missing. Instead of the colours being a, ab, bc, c, the flower will now have colours a, a, c, c. Since colour a alone corresponds to sepal identity, and c alone signifies carpel identity, a flower with rings a, a, c, c will have sepals in the outer two whorls and carpels in the inner two, giving the formula sepal, sepal, carpel, carpel. This is essentially what the mutant flowers belonging to the b class look like. The model has been expressly designed to account for the b class of mutants in terms of the loss of a particular hidden colour: b.

The a and c classes of mutants can be explained in a similar manner, through loss of their respective colours. In this case, though, there is an additional complication. To predict the correct pattern of organ identities, we must assume that the a and c colours are not completely independent but oppose each other in some way. If for some reason colour a is missing, then the c colour appears in its place. Similarly, if c is missing, the a colour will substitute. Thus, in a mutant that lacks a, the c colour appears in all rings but the b colour is not affected, giving the colours c, bc, bc, c. This would signify a flower with the formula carpel, stamen, stamen, carpel, agreeing with the appearance of class a mutants. On the other hand, if we take c away, the a colour appears everywhere and we get a, ab, ab, a, signifying a flower that is sepal, petal, petal, sepal, as observed with class c mutants. These rules may seem rather arbitrary, but remember that at this stage they have simply been devised to account for the appearance of the mutants. We shall return to how hidden colours can actually oppose each other in a later chapter.

The model therefore gives us a set of rules for predicting what type of organs will be made when a distinctive regional quality, symbolized by a colour, is lost. We can even predict what would happen if two hidden colours were missing. Suppose both colours b and c are absent: the flower would only be left with a, and because there is no c to oppose it, a will appear in all rings, predicting a flower that only consists of sepals. This is precisely what is seen when class b and c mutations are combined in the same plant.

Identity Genes

The effects of hidden colours in a rather negative sense, by showing what happens when they are removed. This is because of the reverse way in which we learn the DNA language through *mutations*, looking at what happens when a particular gene is defective. From a positive viewpoint, we could say that there are a specific set of genes in the plant, what *organ identity genes*, that are dedicated to producing the set of a, b and c colours. The positive significance of these genes is to ensure that particular colours are made. *Mutations* that render one of these genes ineffective result in the loss of a colour, and so change the identity of the whorls of organs that develop.

It is important to emphasize that neither these genes nor the colours they produce represent instructions for how to construct a particular type of organ. They simply provide distinctions between regions. It might be thought, for example, that because a + b results in an organ developing with the identity of a petal, then this colour combination specifies how a petal should be made. To see why this is not the case, which compares flowers from *Antirrhinum* with *Arabidopsis*. The basic organization oft he two types of flower is the same: they both consist of concentric whorls of sepals, petals, stamens and carpels. This reflects a similar distribution of a, b and c hidden colours in concentric rings. Nevertheless, the structure of the various organs is quite different, allowing us to distinguish the two species quite easily. For one thing, the *Antirrhinum* organs are much larger, being about

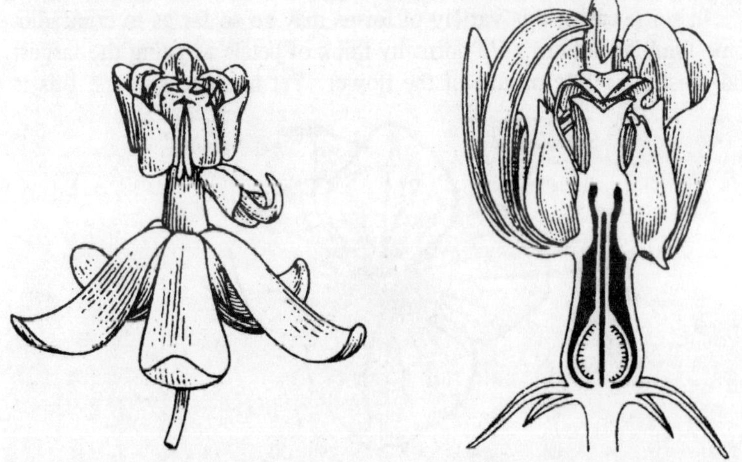

Fig. 11.6. Left—Asclepias curassavica; Right—Asclepias curassavica (vertical section.

ten times the size of *Arabidopsis* in the linear dimension. But even adjusting for size, the organs obviously have a different structure. The five petals of an *Antirrhinum* flower are united together for part of their length to form a tube. At the end of the tube, the petals are more separate, forming five lobes, the lower ones providing a platform for bees to land on and prise open the flower. In contrast, the petals of *Arabidopsis* are more spoon-shaped and are entirely separate from each other. Together, they form a symmetrical cross (hence the name Cruciferae, for the family of plants this species belongs to). Similar comparisons could be made for the sepals, stamens and carpels: in each case there are numerous differences in anatomy and shape that distinguish corresponding organs of *Antirrhinum* from *Arabidopsis*. So even though the identity of the organs in both species depends on a similar set of hidden colours, the detailed structure of the organs is different.

The point is that if the a, b and c hidden colours were giving precise instructions on how to make each type of organ, the organs should be identical in both species. If the details of how to make a petal were specified by the a + b combination, a petal of Antirrhinum should look the same as one from Arabidopsis. Clearly the colours are not giving instructions of this sort. They merely provide a distinction between different regions, allowing organs with separate identities to develop. It is as if the colours provide a common underlying pattern, but how this becomes manifested in the final organs of a flower can vary greatly according to the species.

In some cases, this variety of forms may go so far as to contradict some familiar notions. We normally think of petals as being the largest and most attractive organs of the flower. Yet in some species, this is

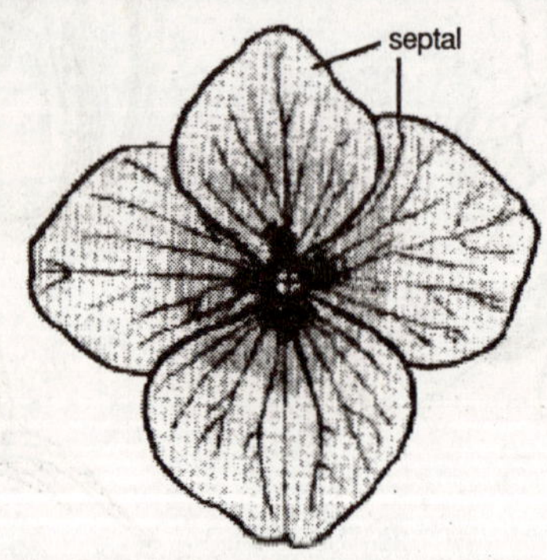

Fig. 11.7. Flower of Hydrangea with large showy septals.

a feature of the outer whorl of organs, the sepals rather than the petals. In flowers of the genus *Hydrangea*, for example, the sepals are sometimes much more conspicuous than the petals, so the colourful display we enjoy in garden varieties is almost entirely due to the sepals. Although the relevant genes from these species have yet to be studied, it is reasonable to suppose that they will have a comparable set of a, b and c hidden colours to those in *Antirrhinum* or *Arabidopsis*.

It is just that in the case of *Hydrangea*, this pattern of colours becomes manifested in a different way.

In the previous chapter, we also came across the notion of hidden colours that can be variously manifested. Recall that they were invoked as a way of accounting for the pattern of bristles in a fly. The fly was assumed to contain a patchwork of hidden colours, a set of regional differences that could not be seen. Although normally invisible, the patchwork could nevertheless be inferred from the way it was interpreted, resulting in one particular bristle pattern being displayed out of many other possibilities; just as someone flying over a football stadium and seeing various messages might infer that the people down below were responding to a grid of colours on the ground. The hidden colours that make up this patchwork are similar in kind to those have already been described in this chapter as a, b and c. It is just that in the flower example, the patchwork has the configuration of four concentric rings.

In the previous chapter, a pattern of hidden colours was inferred from mutations that changed the way they were interpreted by a *particular gene* (scute). The hidden colours were left unchanged by the mutations; all that was affected was their interpretation. In this chapter, it is the hidden colours themselves that are proposed to change in order to account for the appearance of various mutants. In this case, the mutations are affecting a specific set of genes, the organ *identity genes*, needed for producing the a, b and c hidden colours. Each mutation results in the loss of a colour, changing the identity of the organs in some way. So in one case the mutations affect the interpretation of the hidden colours; whereas in the other case the mutations influence the production of the hidden colours. In both cases, the hidden colours provide a frame of reference that can be variously interpreted, eventually becoming manifested in the visible structure of the organism that develops; the colours are not providing a set of instructions for how to construct a particular organ or a bristle pattern. All this discussion of hidden colours and their interpretation may sound rather abstract at this stage, but I would ask the reader to be patient and eventually their meaning and utility for understanding development will become dear.

A Change in Outlook

Looking back on Goethe's views from our present perspective, we can see that many of his ideas turned out to be penetrating. The idea that the different organs of a plant might be variations on a theme has a modern resonance with the various hidden colours that confer distinct organ identities. In my view, though, Goethe's greatest insight was his clear perception of how the study of abnormalities, what we now call mutants, could be used to understand the normal course of development. As he stated in his essay on plant metamorphosis:

From our acquaintance with this abnormal metamorphosis, we are enabled to unveil the secrets that normal metamorphosis conceals from us, and to see distinctly what, from the regular course of development, we can only infer. And it is by this procedure that we hope to achieve most surely the end which we have in view.

He dearly saw that this reverse form of logic, arguing from the abnormal to the normal, was a valid and important way to proceed in unravelling development. Perhaps it was Goethe's breadth of mind, his desire to understand the underlying unity of nature without too much concern for experimental details, that led him to this remarkable insight. Everything Goethe said about plants was gospel. Some of his ideas, like his notion that organs change in appearance due to a sap being gradually purified as plants develop, are of little modern significance. But his clear appreciation of the significance of abnormalities was certainly ahead of its time.

Goethe's perspective only came to experimental fruition in the twentieth century, as mutations affecting development started to be investigated in detail. The unravelling of the abc model is a good example of how the outlook underwent a change. Given its basic simplicity, it seems quite remarkable that the abc model for flower development was only proposed in the late 1980s, even though the experimental approach that lay behind it, the production and classification of mutants, had been well established for many decades before this. The advance had more to do with a change in the way that flowers were being looked at than in the development of new technology. When we had first obtained one of the class a mutants (carpel, stamen, stamen, carpel), going home in the evening after having spent some time looking at its flowers. It was clear that the outer whorl of sepals had been replaced by female organs, but it was less obvious what had happened to the next whorl, where petals normally form. It seemed that these organs were narrow and strap-like with abnormal structures at the ends. Various models at home, it occurred to me that if the strange strap-like structures were due to a transformation of petals towards male organs, the stamens, a simple model could account for the various classes of mutant we knew about. The next morning, we rushed into the greenhouse to look at the mutant flowers again. To my delight the strap-like organs did indeed have some tell-tale features of stamens overlooked the previous day. Later on we obtained some much clearer examples of this type of mutation where there could be little doubt that stamens had replaced petals, but the earlier anticipation of the result has remained with me as a striking example of how observations and descriptions are influenced by what you are looking for. In the 1980s we had started to look at flowers in a different way. At the back of our mind we had the notion that genes might act in combination to confer distinctions in identity. And one of the most important contributions to this new outlook on flowers came from studies on a quite different organism: the fruit fly, *Drosophila melanogaster*.

Segment Identity in Flies

About one hundred years after Goethe wrote his treatise on plants, the zoologist William Bateson described a comparable set of abnormalities in the animal world. Bateson was convinced that the only way to understand evolution was by studying how biological forms vary, and he therefore set about cataloguing the principle types of variation in his book Materials for the Study of Variation of 1894 (Bateson was later to become one of the founders of the science of genetics, a term he coined). Among the variations he described were some striking abnormalities in insects and crustaceans. Like plants, these animals are divided into parts that seem to be fundamentally equivalent. Look at a shrimp or a fly and you can easily see that they are made up of repeating units or segments. Almost the entire body of these animals seems to be based on segments that can be modified in various ways: some bear legs, some can have wings, while others have no appendages. Even the head, which bears antennae, eyes and various mouthparts, can be considered as being made up of segments.

Bateson came across abnormalities in which part of a segment seemed to be converted or transformed into something typical of a quite different segment: an insect with a leg at a position normally occupied by an antenna, or a crab with an antenna instead of an eye. It seemed that one sort of appendage was replacing another What was surprising in all of these cases, just like in those of the flower, was that you didn't end up with a complete mess. The extra antennae or legs seemed to be remarkably normal even though they were growing in the wrong place. Bateson realized that these abnormalities were of great significance: 'Facts of this kind, so common in flowering plants, but in their higher manifestations so rare in animals, hold a place in the study of Variation comparable perhaps with that which the phenomena of the prism held in the study of the nature of Light' He coined a special term, homeosis, for this particular type of variation, in which one member of a repeating series assumes features that are normally associated with a different member

Yet, in spite of Bateson's prophetic words, the real significance of these homeotic mutants was overlooked for a very long time: they were treated as no more than curiosities. Strangely enough, it was their dramatic effects that discounted them in most people's eyes. Most of us have had the experience of desperately kicking or banging a machine after it has refused to perform properly. Occasionally, such acts of desperation can jolt the machine into working again but, although we may be pleased with the outcome, we do not imagine that we understand the machine any better after this. If anything, it seems even more mysterious to us when it responds to acts of frustration. This was the early view of homeotic mutations: that they were genetic jolts that caused a major change in development but were not themselves informative about the underlying mechanisms. Development was thought to be so complicated and subtle that major flips of this type were unlikely to be revealing. It was only during the 1960s and 70s, with the work of Ed Lewis at Caltech on segments in fruit flies, that the central importance of these homeotic changes began to emerge.

Before explaining the results Lewis obtained, briefly to describe a normal fruit fly. We can consider the main body of the fruit fly as being made of 14 segments, numbered starting from the head end. The *head* contains three segments (0, 1 and 2). The next region of the body, called the *thorax*, comprises another three segments (3, 4 and 5), each of which carries a pair of legs. Segment 4 also bears a pair of wings whilst segment 5 has a pair of small appendages, called halteres, which are thought to help balance the fly during flight. The rest of the fly, the *abdomen*, is made of eight segments (6-13). In the 1940s, Lewis had been searching for mutants that might be useful for

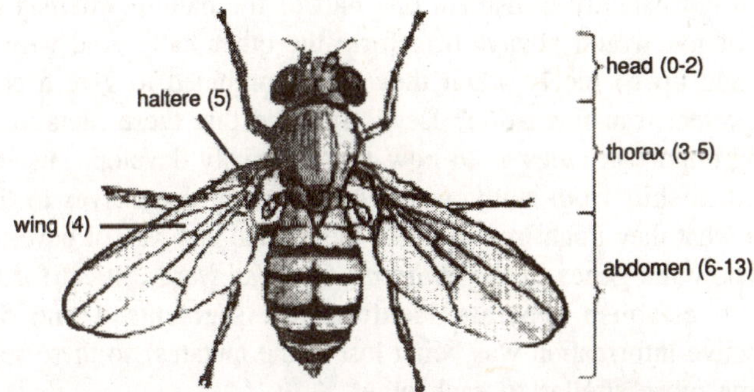

Fig. 11.8. Adult fruit fly, with segment numbers indicated in brackets.

studying the structure of genes, and he came across a mutant called bithorax. In the *bithorax* mutant, the *halteres*, the tiny balancing organs on segment 5, seemed to have been partly transformed into wings. It was as if segment 5 had become more like segment 4, where wings normally develop. As Lewis worked intensively on bithorax mutations, he started to realize that he was dealing with a complex of several genes, arranged next to each other along the DNA chromosome. Every gene corresponds to a small stretch of DNA, so in this case Lewis was dealing with several of these short stretches, lined up one after the other. He called this duster of genes the *Bithorax Complex*.

Lewis discovered that many of the mutations in the *Bithorax Complex* affected only part of the haltere. The original bithorax mutations, for example, appeared to transform only the front half of the haltere into the corresponding half of the wing. Another type of mutation gave the complementary result: only the rear half of the haltere was transformed into rear wing. As he recalled: 'You see, bithorax mutants were said to transform the haltere into wing. That was wrong. They only transformed the anterior part. So it was very exciting to find a mutant that did the complementary thing' It was as if you could transform different parts of the haltere separately.

He then wanted to see whether the effects of the two sorts of mutation could be added together, to simultaneously transform both the front and rear halves of the haltere into wing. To test this, he crossed flies with the different types of mutation with each other, and eventually managed to combine both mutations in the same individual fly. The two types of transformations did indeed add up precisely, to give halteres entirely replaced by wings, resulting in a four-winged fly. This was a key advance because it told him that these mutations could not simply be ignorant kicks of the system. Why would one type of jolt consistently transform one half of the haltere whereas another type of jolt would always transform the other half? And why should they add up so nicely when they were combined to give a complete and perfect transformation? Lewis realized that these mutations were giving important clues as to how flies normally develop. His interests started to shift from studying the mutations in themselves to thinking about what they might reveal about the normal pathway of development. Somehow the genes in the *Bithorax Complex* were critical for normal flies to establish separate identities for segments 4 and 5. This distinctive information was being lost in the mutants, so these segments became more similar to each other.

For a long time Lewis continued to collect and study many developmental mutations in different parts of the *Bithorax Complex*, each affecting the adult fly in different ways. Then, in the 1970s, he managed to get a mutation that removed a large chunk of DNA containing all of the genes in the *Bithorax Complex* in one go. How would a fly develop if it had none of the genes in the *Bithorax Complex*? The result was that the mutant died at a very early stage, just about the time that the larva was hatching from the egg case. To understand what was going on, Lewis therefore had to look at the early larvae rather than adult flies.

As with many insects, fruit flies spend their early feeding life as larvae, small grubs that eventually pupate and metamorphose into the adult flies. Larvae can also be divided into 14 segments, corresponding to the adult ones that will form later, but there are no appendages such as wings, legs or halteres to distinguish them. For that reason, larvae were considered to be rather uninformative, and nobody had taken much of an interest in looking at early larval stages of development. When Lewis saw that his deletion mutant died at such an early stage, he was forced to develop a method for cleaning and preparing the young larvae or embryos. A key step was treating these embryonic larvae with a chemical, lactic acid, that made it much easier to see their outer surface or skin, called the cuticle. Using this method, Lewis quickly saw how each segment on a larva could be distinguished from the others by its characteristic pattern of tiny thorn-like outgrowths, called denticles, on its outer surface. Look at the larva on the left which shows most segments of a normal individual identified in this way, according to their specific patterns of denticles.

Having developed this method for looking at larvae, Lewis could then examine the mutant that died early on. As he later recalled:

Being lethal, we were forced to find out what the thing looked like. It actually crawled out of the egg. It was no good just looking at one dead animal. Then used a lactic acid method and it worked beautifully. When you come right down to it much of the success of bithorax was the discovery that you could make simple, quickly prepared mounts of the embryo, in which the cuticle pattern allowed you to read gene function, right off like a book. So often it's a little technique that allows you to make a big jump that was what really helped because we could study any mutation by putting a larva in a drop of lactic acid and alcohol.

By looking at the pattern of larval segments in his mutant with the entire *Bithorax Complex* deleted, Lewis made a remarkable discovery. The segments from 5-13 were no longer distinct from each other: they all resembled segment 4. In other words, the *larva* was normal at the head end (segments 0-4), but then it had nine rear segments that all looked like segment 4. Remember that in the mutants, it was only segment 5, the region that bears the halteres, that was transformed to resemble segment 4. But in the case of this new mutant, all of the segments from 5 onwards had assumed the same identity as segment 4. If such a larva could have survived to adulthood, it would have turned into a fly with 22 legs, the same number as an entire soccer team (it would also have had 20 wings).

Lewis went on to produce further deletions, removing only some of the genes in the *Bithorax Complex*. These gave similar types of result. For example, in one case the head and middle segments (0-6) were normal, but then the remaining segments behind this all looked like segment 6. A general pattern was emerging from these mutants. The larvae might be normal from the head up to a certain point, segment 6 in this case, but then the segments behind this would simply reiterate the same identity.

To explain his results, Lewis proposed that the normal role of the *Bithorax Complex* was to produce a set of distinct 'substances' in each segment. The combination of substances could lead to each segment developing with a particular identity. The model can get rather complicated, in terms of hidden colours that nevertheless captures the essential features.

In my simplified model, the fly has only four segments, with distinct identities 1. The identities depend on a set of three overlapping basic hidden colours, denoted as f, g and h, distributed so that there are progressively more colours as you move towards the tail end of

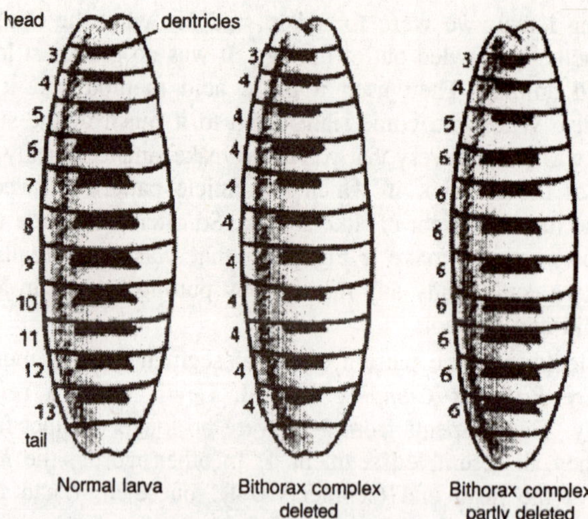

Fig. 11.9. Larvae from normal fly (left), mutant with all genes in the Bithorax complex deleted (middle), mutant with part of the Bithorax complex deleted (right).

the normal fly: f is in three segments (2, 3, 4), g is in two segments (3, 4) and h is only in segment 4. The identity of a segment depends on the combination of hidden colours it contains. So if we remove the rear-most colour h, the identities of the first three segments would be unchanged but the fourth segment would now only have f and g, giving it identity 3. Removing g as well as h would leave only fin three segments, giving them all identity 2 (*mutant hg*). Removing all three colours would give all segments identity 1 (*mutant hgf*). In other words, with this model we can account for the various mutants Lewis observed that were normal up to one point, but then reiterated the same type of segment, by proposing the loss of one or more hidden colours.

This is essentially how the genes in the *Bithorax Complex* confer distinctions between segments. We can say that the *Bithorax Complex* contains a set of segment identity genes. Each of these genes is needed for a particular hidden colour, and the combination of colours determines the particular identity of the segment. The normal role of these genes is therefore to generate hidden colours in their respective regions. Mutations that inactivate or remove one or more of the segment identity genes result in a loss of colour, and hence a change in some of the identities. (In the real fly, the genes in the *Bithorax Complex* have to discriminate between the nine segments from 5-13. You might expect, as Lewis did, that this would involve eight separate substances or colours. However, it appears that only three major colours are involved as some of the differences between segments 5-13 are due to variation in the intensity of colour as well as the type of colour.)

Although the genes in the Bithorax Complex confer distinctions between segments 5-13, we still have to explain how the segments nearer to the head (0-4) get their distinctive identities. It turns out that there is another cluster of segment identity genes, called the *Antennapedia Complex* (Antennapedia is so named because a mutation in one of its genes results in legs growing in place of antennae). These genes act in a similar way to those of the *Bithorax Complex*, providing a further set of five hidden colours (which we can call a, b, c, d and e) that distinguish the segments at the head end (the role of the a and b colours is more restricted than the others). From head to tail, the distinctive segment identities in a larva, and the fly that develops from it, therefore depend on the combined action of eight hidden colours a-h. The first five, a-e, depend on identity genes in the *Antennapedia Complex*, whereas the last three, f-h, depend on those in the *Bithorax Complex*.

As with previous examples, these hidden colours are not instructions on how to make a particular type of structure, such as a segment bearing legs and wings; they simply provide distinctions between regions. For instance, a similar set of hidden colours to those in fruit flies has now been documented in many different types of insect, including beetle and butterfly species. Even though these species look very different outwardly, they have a common set of underlying hidden colours that provide distinctions between their segments from head to tail. Many of the obvious differences between these species most likely have to do with the variety of ways in which this common pattern can be interpreted and becomes visibly manifest. The hidden colours do not represent detailed instructions, they just provide a frame of reference.

Flowers and Flies

Monstrosities in flowers and flies, in which parts of the organism appear to adopt mistaken identities, can be explained in very similar ways. In each case we can infer a pattern of hidden colours that provide distinctions between different regions of the organism, formalized as concentric rings in flowers or a linear array in flies. These regional differences in colour can be interpreted to give organs or segments with distinct features. The colours themselves depend on a particular set of genes, called organ or segment identity genes. Take away or mutate one of these genes and colours are lost from the pattern, changing the identities of some of the structures that develop. These examples have been chosen from plants and animals to show how hidden colours depend on genes. In the previous chapter it was shown how genes are also involved in interpreting these colours, responding to them in an informed way. How hidden colours are produced and how genes are able to interpret them. To answer this satisfactorily, we have to know more about the molecular properties of the genes and proteins involved. We have to turn from what these genes signify to how they are used.

12

INTELLIGENT CREATOR

During the nineteenth century, theologians and biblical scholars became engaged in serious study of the antecedents of the *Bible* as we know it, comparing the various ancient texts from which it evolved, seeking to understand the apparent contradictions, and attempting to identify those texts that might have been closest to the original. While these scholars were busy studying "the Word" to illuminate God's great act of creation, other pious Christians were studying "the Work" itself—the many species of living creatures—in order to worship God better through a deeper appreciation of His handiwork. These religious students of nature called themselves natural theologians.

Natural theology was of considerable interest to clergymen and others in England. It was based on the belief that since all life had been created by God, a detailed study of plants and especially animals could reveal some aspects of the mind of the Deity. Clergyman John Ray had argued just that point in 1691 in The Wisdom of God Manifested in the Works of the Creation. More than a century later another English clergyman, William Paley, archdeacon of Carlisle, published Natural Theology: or Evidences of the Existence and Attributes of the Deity Collected from the Appearances of Nature (1802). It included a great deal of data on the anatomy of organisms and their remarkable adaptations to the environment in which they lived. Paley emphasized that life was so complex that it could only be the result of a divine power— such a complex design must have a designer. How could one possibly explain a single organism or even one of its parts, such as the eye, without invoking an "*intelligent Creator*"? It was impossible to imagine that a human being or an eye could "*just happen.*" This notion has had a rebirth recently as I.D., or intelligent design.

In the 1830s still another clergyman, the Right Honourable and Reverend Francis Henry, earl of Bridgewater, left a large bequest for the publication of a series of volumes "On the Power, Wisdom, and Goodness of God, as manifested in the Creation; illustrating such work by all reasonable arguments, as for instance the variety and formation of God's creatures in the animal, vegetable, and mineral kingdoms; the effect of digestion, and thereby of conversion; the construction of the hand of man, and the infinite variety of other arguments; as also by discoveries ancient and modern, in arts, sciences, and the whole extent of literature." Eight of these Bridgewater treatises were written by well-known scientists and published. They dealt with astronomy, physics, meteorology, chemistry, mineralogy, geology, and biology. They organized the available scientific knowledge to show what God had accomplished and conversely, suggested that science made sense only in relation to the Deity.

The natural theologians thought of their work as parallel to that of the bookish theologians. Studying God's Work in order to better understand Him seemed a reasonable approach in the eighteenth and early nineteenth centuries, when serious questions were being raised about the Bible's accuracy. Some natural theologians even went so far as to argue that the statements of God, whether transmitted in oral or written form, were surely subject to greater errors at the hands of humans than God's visible handiwork in living nature. The Word might err, but the Work spoke the truth.

In one of the most ironic episodes in intellectual history, the vast body of information the natural theologians accumulated about the life histories of plants and animals constituted much of the database Charles Darwin would draw on to support his concept of evolution. The beautiful adaptations of plants and animals could not be denied; all that was required was to switch the explanatory hypothesis from divine will to natural causes. Whereas the natural theologians began with the answer—divine creation—and then used the data they had gathered from nature to support the answer they had already decided was true, Darwin began with the data of adaptation and followed them wherever they led.

Charles Robert Darwin (1809–1882) was the son of a well-to-do English physician. He began and then abandoned a career first in medicine and then in the Church. From childhood, Darwin's first love had been nature—he had a particular fondness for beetles—and for several years he sought an explanation for many of the puzzles that he, along with the natural theologians, had already observed in the world around him. In addition to the amazing adaptations of plants and animals to their environment, he noted the sequential change of fossils in the geological record, the seemingly hierarchical relationships of organisms from simple to complex, some surprising discoveries by embryologists about early development, and the similarity of species living geographically close to one another. To understand why these items puzzled Darwin and his contemporaries, and why evolution seemed to offer the key, we must travel backward in time momentarily to the Scientific Revolution.

Paving the Way for Evolution

Among the great scientific curiosities during the two centuries before Darwin's time were objects called "figur'd stones" or, as Darwin's contemporaries would call them, fossils. Some of the figures in these curious stones were very similar to the oysters, clams, and snails that were common along the sea coasts of the world, yet the stones were often found high up on mountains. Perhaps they had been deposited there during Noah's flood. But still, how could an oyster get inside a rock?

Robert Hooke (1635–1703), a prominent member of the Royal Society of London and best known to biologists as the discoverer of cells, was much interested in fossils. He and his contemporaries collected them and, by publishing and comparing their findings, developed the skills to distinguish what had originated from a living creature from what had not. They came to the conclusion that fossils were formed when an organism died in the sea or a lake, sank to the bottom, and was covered by silt. As the silt increased in depth, the pressure would slowly convert the buried layers of silt and the entombed creature to stone. During this process, the original material—bone, shell, or a portion of a plant—would itself turn to stone. That is, the original material would be replaced, with almost molecule-for-molecule exactness, by the chemicals in the silt. This process of petrification solved the puzzle of how an oyster could get inside a rock.

But other puzzles were more difficult to solve. While some fossils seemed to be identical to living species such as shellfish, for most fossils there was no known living representative. This presented the natural theologians with a serious problem, because the Bible, at Ecclesiastes 4:14, says: "I know that whatever God does endures for ever; nothing can be added to it, nor anything taken from it." This teaching was widely accepted to mean that there have been essentially no significant changes in any organisms since creation. Minor variations might occur in varieties and breeds, but race horses and draft horses were still horses and all varieties of roses were still roses. The fossil findings, however, suggested that some creatures had failed to endure forever—they had become extinct.

The natural theologians came up with a temporary answer. Though there was no living mollusk on the coast of England that resembled the fossil species found there, individuals of the species could still be living elsewhere, possibly off the coast of Africa, and had simply not been discovered yet. A few such examples had already been found. But as more and more observations were made, it became highly probable that some fossils represented species that had become totally extinct. This was difficult to accept since it implied that God had made a mistake.

Two major types of rocks, volcanic and sedimentary, were recognized by geologists in Darwin's day. Volcanic rocks are formed by lava and ash from volcanic eruptions that became consolidated into rocks as they cooled. *Volcanic rocks* almost never contain fossils. In *sedimentary*, or stratified, rocks, on the other hand, the sediments are deposited in layers or strata, with the most recent stratum deposited on top. *Fossils* are found in sedimentary rocks, often in great abundance. By the nineteenth century it was clear that a given layer's position in sedimentary rocks provides valuable information about its age relative to the other layers—and hence about the age of any fossils it contains. Each layer of rock is older than the one above it and younger than the one below it, in the same way that the garbage at the bottom of a dumpsite is older than the garbage on top.

Most sedimentary rocks are formed at the bottom of oceans, in inland seas, or on continental shelves, where silt is washed down by rivers from higher elevations. We now know that at different times in the past much of the interior of the United States was covered by shallow inland seas. Each time the sea was present, sedimentary rocks would form at its bottom and preserve fossils of species prevalent during that time period. When the seas receded, the newly formed strata would be exposed. When subsequent changes in the Earth's crust elevated the strata in one area and exposed them to the surface, they became visible—and available for study by geologists.

Early in the nineteenth century geologists began to classify exposed rocks on the basis of their mineral composition and especially on the basis of the fossils they contained. The first major study of this sort was done in England by William Smith and published in 1815. He studied the strata in cliffs in numerous localities with the goal of arranging them in the order of their deposition. At one site he might recognize strata A, B, C, D, and E from bottom to top. At another site he might recognize F, G, H, I, and J. However, close study might show that F and G were identical to D and E, both in composition and in the fossils they contained. Thus he could conclude that the proper sequence, from oldest to youngest, was A, B, C, D F, E G, H, I, J.

Over the decades, this procedure produced the worldwide "*geological column*," an imaginary pile of strata that covers the span from the oldest discovered sedimentary rocks to those formed most recently. The height of this column—that is, the thickness of all known strata added together—is estimated to be about 60 miles (over 100 kilometers). This does not mean that one can start digging at any place on the Earth and go through 60 miles of strata. This impressive height is based on adding up all the different strata in the various places where they occur. How long did it take for all these strata to form? No one in the early nineteenth century knew, but geologists realized that it must have been a very long time, because silt washes into inland seas very slowly.

Continued studies established that each major group of strata contains its own unique kinds of organisms. The famous French naturalist Georges Cuvier (1769–1832) interpreted these data as evidence that at various times in the past drastic catastrophes had destroyed all life, and subsequently there had been new creations of quite different species. While Cuvier's theory, called catastrophism, held on to the idea of a divine creator for every single species, it significantly modified the creation process outlined in *Genesis*. Instead of taking just one week, creation in Cuvier's theory stretched throughout the entire history of life. And instead of recognizing just one great catastrophic flood, Cuvier suggested that life-destroying catastrophes had occurred over and over again.

An alternative explanation to catastrophism was evolution—the gradual change of species into other species over time. Darwin was not the first person to think of evolution; the concept had been around for centuries. Even the classical Greeks had speculated along these lines but then abandoned the idea when no data could be offered in support. The new observations and speculations of the seventeenth and eighteenth centuries, however, slowly laid the path toward a workable evolutionary theory.

One concept that helped pave the way was the scala naturae, or scale of nature—the suggestion that all animals could be arranged in a linear series based on increasing complexity, with no appreciable gaps in the series—from amoeba to humans. Where gaps seemed to exist, there were presumably intermediate forms yet to be discovered. Thus the great apes seemed to link human beings with other mammals, seals and whales linked fishes with land-living vertebrates, and bats were considered intermediate between birds and mammals. The roots of this concept could be traced back through medieval times to the Greeks, and it was still widely accepted in Darwin's day.

A further observation that prepared the way for evolution was that species of animals and plants are not randomly different from one another but seem to fall into naturally hierarchical groups. Similar individuals can be classified as the same species, similar species can be included in the same genus, similar genera in the same family, similar families in the same order, similar orders in the same class, similar classes in the same phylum, and similar phyla in the same kingdom. The first systematic attempt to classify living nature in this manner was made by the Swedish naturalist Carolus Linnaeus in the eighteenth century. In time, both the scale of nature and hierarchical classification were understood in terms of evolution—similar groups, such as species within a genus, are alike because they descended from a common ancestor. At the next level in the hierarchy, all the species of a genus of birds and indeed all species of birds, are descended from a very ancient common ancestor.

The person who first tried to bring ideas about evolution together into a coherent theory was the Frenchman Jean Baptiste Lamarck (1744–1829), who in his Philosophie zoologique (1809) maintained that one species evolves into another species in order to better adapt to its environment. Observing fossils in France, Lamarck noted that one geological stratum might have an abundance of one species of mollusk with little variation. The next higher stratum might contain species that were similar, but none would be exactly like those in the lower stratum. As he studied progressively higher strata, Lamarck observed that species became steadily different over time, stratum by stratum. Since fossils in a lower stratum were known to be geologically older than those in a higher stratum, it stood to reason that though a fossil in a higher stratum could not be the ancestor of one in a lower stratum—descendants cannot live before ancestors—a species in the lower stratum just might be the ancestor of a species higher up in the column. Lamarck concluded that what he was seeing in the fossils of progressively higher strata was change in a lineage over time. This hypothesis was markedly different from Cuvier's view that as the species in one stratum became extinct, closely similar ones were created anew and preserved in the next higher stratum.

Lamarck postulated a changing environment as the mechanism for the evolutionary change he observed. Species evolved in order to adapt, he believed. His classic example was the giraffe's remarkably long neck. The ancestors of today's giraffes, he said, had short necks and grazed on grasses and low shrubs, as do most other herbivorous mammals. Lamarck suggested that some ancestors of modern giraffes attempted to exploit a new and abundant food source—the higher leaves of trees. To reach the leaves they had to stretch their necks, which gradually lengthened with so much stretching. Lamarck thought that traits that came about through repeated use could be passed along to offspring. Thus giraffes would inherit the long necks of their parents and then stretch their own necks even further; over many generations, giraffe necks would become longer and longer until they reached the length of giraffe necks we see today. Conversely, characteristics that were not used would eventually wither away, as happened to eyesight in moles and bats.

This hypothesis of evolutionary change through "the inheritance of *acquired characters*" (meaning "*characteristics*") was not widely accepted in the early nineteenth century, since it was contrary to the Bible and was based on too much speculation and too few data. Other people besides Lamarck, including Charles Darwin's own grandfather Erasmus Darwin (1731–1802) had suggested that evolution might occur, but no one had yet argued the case well enough to convince the scientific community. Thus, in the first half of the nineteenth century the dominant scientific position was that species are "*fixed*," that is, they do not evolve. Although questions about the accuracy of the *Genesis* account of creation were being asked by scientists as well as biblical scholars, and alternative scientific as well as theological interpretations were being offered, in Charles Darwin's day none of these theories was taken seriously enough to undermine the traditional Judeo-Christian teaching. Evolution was out of favour; divine creation was still in vogue.

Nevertheless, in 1858 when Darwin sat down to prepare his treatise on evolution for publication, rich veins of biological and geological data were just waiting to be mined in evolution's support. Collections of animals and plants were being assembled and preserved in museums and herbaria at a prodigious rate—Cuvier himself had put together an outstanding zoological collection, especially of bones. Thousands of new species had been carefully described by taxonomists and placed in the Linnean system of classification, primarily on the basis of structural similarity. The embryonic development of the major types of animals was beginning to be understood. The main areas yet to be well understood were ecology, behaviour, physiology at both the cellular and organ level, the study of microorganisms, and especially inheritance. Ignorance of the mechanisms of inheritance—or as we would now say, genetics—was to prove a serious problem for Darwin.

The geological data Darwin would draw on to support his theory of evolution had been well summarized in Charles Lyell's Principles of Geology (1830–1833). Lyell, an Englishman, saw no geological evidence for the vast catastrophes that Cuvier had invoked to explain the sequence of unique fossil faunas in strata of different ages. Lyell believed that all the usual geological events–volcanic eruptions, earthquakes, erosion by wind and water, and the uplift of land—were adequate to explain former changes in the Earth's surface, just as they explain them today. Thus, Lyell was a uniformitarian—a supporter of gradual geological change—not a catastrophist like Cuvier. Darwin, convinced by the arguments of his friend and countryman Lyell, would also favour gradual change over time and would reject sudden leaps in evolution.

Beyond these advances in biology and geology, science in general was becoming better organized by the middle of the nineteenth century. Publications, the life blood of any science, were numerous and of high quality, and biology and geology were being taught in universities. The Western world was in prime shape for a paradigm shift.

Origin of Species

Charles Darwin did not set out to prove evolution. As a young naturalist, he thought he knew, as did all his contemporaries, that species are fixed to the degree that one species does not change into another. He knew, of course, that domestic species could be selected to produce strikingly different varieties. Horses could be selectively bred for speed or for strength. Roses could be selectively bred to climb higher, be more beautiful, or smell sweeter. Dogs and even pigeons could be bred to exhibit all sorts of new shapes and behaviours. But wild species of plants and animals were thought to be uniform, despite minor variations in individuals, and to remain essentially unchanged for centuries. For example, the ancient Egyptians embalmed animals of many species. When these were examined closely in the nineteenth century, three thousand years later, they seemed to be identical with contemporary individuals. These data were especially important in buttressing the *Genesis* account, since it was assumed that the early Egyptians lived not long after creation itself.

During the 1830s, when the Bridgewater treatises and Lyell's Principles of Geology were being published, Darwin embarked on the formative event of his scientific life—a voyage around the world on the *H.M.S. Beagle*, with Captain Robert Fitzroy in command. The purpose of the voyage was to map coastlines, for reliable navigational charts, necessary to avoid shipwrecks, were unavailable for most coastlines at the time. The young Darwin went along as the ship's naturalist, a common role in those days. It was on this voyage that he made some very puzzling observations that suggested that species might not be "*fixed*" after all. Fortunately, he left a written record of what had set his thoughts along this new track.

While the Beagle was making accurate navigation charts of the South American coast—a slow process with the instruments then at hand—considerable time was available for Darwin to go on shore to observe and collect living and fossil species. During these forays he made two sorts of intriguing observations. One had to do with the armadillos living on the Argentine pampas and a fossil armadillo, the glyptodont. These two sorts of mammals had many features in common, and Darwin assumed they were related. Both the living armadillos and the fossil armadillo-like glyptodonts were found only in the New World, mainly in South America. Darwin's inquisitive mind was always seeking answers, and he might have thought, "Is it not surprising that two such similar forms should have been created in precisely the same part of the world? Could the extinct species perhaps be the progenitor of the living?"

The second class of observations had to do with geographic variations within a species, a phenomenon Darwin first noticed on the mainland of South America. He observed that individuals of what were clearly the same species might vary from locality to locality—and the greater the distance between the localities, the greater the differences. But the most dramatic example of geographic variation Darwin observed was in the Galapagos, a cluster of small islands off the west coast of South America. Most of the species he found there were new to science and restricted to the Galapagos, although they were similar to species on the mainland of South America, 600 miles to the east. Some species were restricted to one island, although a similar species might occur on an adjacent island. Darwin found it astonishing that these islands, so close that most of them were in sight of each other, formed of the same kinds of rocks, with a similar climate, would have recognizably different varieties. Another dramatic example of geographic variation on a small scale was the *Galapagos tortoises* on islands close to one another. Minor differences in structure were such that the local inhabitants could tell the island of origin of any individual tortoise.

The hypothesis that Darwin proposed to account for the armadillo-like fossils and the *Galapagos finches* and tortoises was evolution. Darwin revived this generally moribund notion by suggesting a plausible mechanism for how it could occur. His argument went somewhat as follows: First, the environment on Earth is fixed in size and resources; there is only so much land and sea where organisms can live, find food, and reproduce. Second, every species has the potential for increasing its population size far above a level that the fixed environment can support. A single oak tree in a forest can produce thousands of acorns every year, yet in a mature forest no new tree can reach maturity unless an old tree dies and leaves a space. These two points were well known but by themselves they did not add up to evolution; those few lucky acorns that became trees would be the same kinds of trees as before. Evolution requires change, not a repetition of the same kinds of individuals from generation to generation. Something more was needed to bring about change.

The missing factors were *variation* and *natural selection*. Darwin had observed that individuals of a species vary slightly, that some variants are better adapted for surviving and producing offspring in a given environment than others, and that variations can be transmitted to offspring. In Darwin's day there was little solid information about how inheritance works, but animal and plant breeders knew very well that parents transmitted "*something*" to their offspring that influenced the characteristics of the next generation. They also knew that not all offspring inherited the desirable characteristics to the same degree; even among siblings there was variation, and this difference gave the animal or plant breeder a basis for selecting "*the pick of the litter*" for his breeding stock. Variation and selection rounded out Darwin's theory by providing a mechanism for evolutionary change. The hypothesis in full says:

1. There is neither enough space nor enough resources for all individuals that are born to survive.

2. Some individuals are better able to survive and reproduce under the conditions of a given environment than others.
3. Organisms that survive and reproduce pass along to some of their offspring traits that improve their chances of surviving and reproducing in turn.
4. Over time, this differential survival and reproduction will change a species in ways that better adapt it to its local environment and make it different from closely related species in slightly different environments.
5. Given enough time, the differences between the current generation and its ancestral generation become so great that we say a new "daughter" species has evolved.
6. Similarly, over time the differences between one population and another nearby become so great that we say a new species has evolved.

In making his case for natural selection acting on individual variation, Darwin relied heavily on evidence from plant and animal breeders, whose practices were familiar to him. In the breeders' case, it was human beings, not nature, who chose the characteristics to be preserved by allowing only those individuals with at least the vestiges of the desired features to reproduce generation after generation; all the rest were culled. While the majority of naturalists did not believe that new species in the wild could result from a natural version of this artificial selection, Darwin suspected otherwise.

The Beagle returned to England in 1836, and Darwin proceeded to prepare several books based on his observations and collections, including Zoology of the Voyage of the Beagle and Geological Observations. Neither volume mentioned his developing ideas about evolution. Later he wrote in his autobiography that in 1837 "I opened my first note-book for facts in relation to the Origin of Species, about which I had long reflected, and never ceased working on for the next twenty years" (Barlow 1958). In 1842 and 1844 he wrote drafts of his ideas and told a few close friends, including Lyell and the botanist J. D. Hooker, about the possibility of the evolution of new species. In 1856 Lyell suggested to Darwin that he finish his studies and publish them before someone else anticipated his conclusions. Darwin took that advice and began to prepare a manuscript that was to become On the Origin of Species. The following year he sent a long letter describing his conclusions to the American botanist Asa Gray of Harvard.

Nevertheless Darwin's hypothesis for evolution still remained known to very few people. Various reasons have been suggested for his reluctance to broadcast his ideas more widely. One had to do with his beloved wife, Emma, who would have been upset with a notion so at variance with her deeply held Christian beliefs. One of the Darwin daughters wrote of her mother, "In her youth religion must have largely filled her life, and there is evidence in the papers that she left that it distressed her in her early married life to know that my father did not share her faith" (Barlow 1958). Emma was not the only person upset by the direction Darwin's thoughts were taking. Many educated people in England, scientists and nonscientists alike, who read the Origin when it was finally published found it sorely distressing, since it undermined one of the fundamental beliefs of Western culture. And indeed there is considerable evidence that Darwin's conclusions at first distressed him as well.

Another explanation for Darwin's reluctance to publish is that the concept of evolution was held in low repute; the versions of evolution suggested by Lamarck in France and later by Robert Chambers in Britain had been vehemently rejected by scientists in England. Because Darwin was beginning with a widely rejected notion, it was all the more obligatory that he make an exceedingly strong and well-documented argument. Consequently, a vast amount of reading and thought had to go into the Origin's production—a time-consuming process. Adding to this burden was Darwin's poor health. The medical profession continues to speculate on the nature of his illness to this day; the diagnosis varies from a parasitic disease contracted in South America to psychological problems caused by a domineering father. After Darwin had returned from the Beagle voyage, he and Emma had moved from London to the country—to the village of Down— where Darwin had become almost a recluse by the time the Origin was published. His ill health may have made him reluctant to subject himself to the stress of intense criticism from his scientific peers and others in his social circle.

A final reason for Darwin's slowness was that during the twenty years he thought about evolution before he started the actual writing of the Origin, he wrote other major works based on his findings during the Beagle voyage (two on geology and four on barnacles), and he conducted many scientific experiments to test his various hypotheses. All in all, it was not a bad rate of production for a graying, sickly recluse.

Despite these impediments, Darwin carried through on Lyell's suggestion, and by 1858 the book manuscript was about half complete. Then a bombshell hit. In June of that year Darwin received in the mail a manuscript from Alfred Russel Wallace, an English naturalist who was then collecting in the Malay Archipelago. Unlike Darwin, Wallace came from a modest social background and for many years had made a living by collecting in the New World and the East Indies and selling his specimens to gentlemen collectors and museums. Darwin was thunderstruck with what he read in Wallace's manuscript, as he explained in a letter to Lyell dated June 18, 1858:

My dear Lyell—Some year or so ago you recommended me to read a paper by Wallace in the "*Annals*" which had interested you, and, as I was writing to him, I knew this would please him much, so I told him. He has today sent me the enclosed manuscript, and asked me to forward it to you. It seems to me well worth reading. Your words have come true with a vengeance—that I should be fore-stalled. You said this, when I explained to you here very briefly my views of "*Natural Selection*" depending on the struggle for existence: I never saw a more striking coincidence; if Wallace had my MS. sketch written out in 1842, he could not have made a better short abstract! Please return me the MS., which he does not say he wishes me to publish, but I shall, of course, write and offer to send to any journal. So all my originality, whatever it may amount to, will be smashed, though my book, if it will ever have any value, will not be deteriorated; as all the labour consists in the application of the theory. I hope you will approve of Wallace's sketch, that I might tell him what you say.

Darwin feared he had been scooped. He had developed what was to become the most important theory ever formulated in biology, and before he had made his ideas known, Wallace had reached almost identical conclusions. In science, the rewards go to those who first publish an important discovery or a new theory. Darwin could go ahead and publish first, of course, but if he did, how was Wallace to be given credit for developing closely similar ideas? For Darwin this was a terrible dilemma. Moreover, he was in his usual poor health, and the timing of Wallace's letter proved more difficult for him than it might have to a healthier man. And to augment the anguish, his son Charles, only half a year old, died ten days after Wallace's letter arrived.

Darwin turned to his friends Lyell and Hooker for advice, and they suggested a solution, namely, that they would send both the Wallace and Darwin manuscripts to the Linnean Society of London for joint publication. Darwin agreed, and the two manuscripts were delivered to the society on July 1, 1858, and published shortly thereafter. Part of the letter of transmittal reads as follows:

So highly did Mr. Darwin appreciate the value of the views therein set forth [in Wallace's essay], that he proposed, in a letter to Sir Charles Lyell, to obtain Mr. Wallace's consent to allow the Essay to be published as soon as possible. Of this step we highly approved, provided Mr. Darwin did not withhold from the public, as he was strongly inclined to do (in favour of Mr. Wallace), the memoir which he himself had written on the same subject, and which, as before stated, one of us had perused in 1844, and the contents of which we had both of us been privy to for many years. On representing this to Mr. Darwin, he gave us permission to make what use we thought proper of his memoir, etc.; and in adopting our present course, of presenting it to the Linnean Society, we have explained to him that we are not solely considering the relative claims to priority of himself and his friend, but the interests of science generally; for we feel it to be desirable that views founded on wide deduction from facts, and matured by years of reflection, should constitute at once a goal from which others may start, and that, while the scientific world is waiting for the appearance of Mr. Darwin's complete work, some of the leading results of his labours, as well as those of his able correspondent, should together be laid before the public.

The presentation of the papers by Darwin and Wallace at the Linnean Society and their subsequent publication seemed to arouse little interest. Darwin knew of only one review, and its verdict was "that all that was new in them was false, and what was true was old" (Barlow 1958). In any event, Darwin in great haste prepared a short version of his manuscript, which was published on November 24, 1859, as On the Origin of Species by Means of Natural Selection, or the Preservation of Favoured Races in the Struggle for Life. In contrast with the lack of interest shown at the Linnean Society meeting, the book must have been keenly anticipated because it sold out on the day of publication; a second edition was ready the following month.

Darwin's Reception

The Origin was indeed threatening to most people in the West, for Darwin's arguments could be interpreted as implying a world without either God or purpose. Two Harvard professors, Asa Gray and Louis Agassiz, represent the two poles of reaction among Darwin's most educated readers. Gray was the leading botanist of the time in the United States. Agassiz, a geologist and zoologist, was born in Europe and later came to the United States, where he had a sparkling career. Gray's initial position was similar to that of most mid-century naturalists—orthodox in religious beliefs. But because he had corresponded with Darwin before the publication of the Origin, he was generally aware of the arguments that it would contain, and his review of the book, published in 1860, was careful and fair. He was not fully convinced that Darwin was correct, but he felt that a powerful case had been made for the possibility of evolution and that the matter should be seriously considered:

We are thus, at last, brought to the question; what should happen if the derivation of species [evolution of one species from another] were to be substantiated, either as a true physical theory, or as a sufficient hypothesis? What would come of it? The enquiry is a pertinent one just now. For, of those who agree with us in thinking that Darwin has not established his theory of derivation, many will admit with us that he has rendered a theory of derivation much less improbable than before; that such a theory chimes in with the established doctrines of physical science, and is not unlikely to be largely accepted long before it can be proved. Moreover, the various notions that prevail,—equally among the most and least religious,—as to the relations between natural agencies or phenomena and Efficient Cause, are seemingly more crude, obscure, and discordant than they need be.

The work is a scientific one, rigidly restricted to its direct object; and by its science it must stand or fall. Its aim is, probably not to deny creative intervention in nature,—for the admission of the independent origination of certain types does away with all antecedent improbability of as much intervention as may be required,—but to maintain that *Natural Selection* in explaining the facts, explains also many classes of facts which thousand-fold repeated independent acts of creation do not explain, but leave more mysterious than ever. How far the author has succeeded, the scientific world will in due time be able to pronounce.

Gray's review was widely praised. Darwin regarded it as the best that had been written by that time, even though Gray was far from endorsing the argument in its entirety. He emphasized the problems that Darwin had admitted and pointed out others himself. For example, he noted the critical lack of any real evidence of the origin and nature of genetic variation. Yet in spite of his own ambivalence, Gray insisted on a fair hearing for Darwin, and he became the most vigorous defender of Darwinism in America. He made the prophetic statement that was borne out by subsequent events: the hypothesis "is not unlikely to be largely accepted long before it can be proved."

Gray's principal opponent was his fellow Harvard professor Louis Agassiz, and the debates over Darwinism that were to ensue in America centered on these two individuals. Agassiz, like Gray, was conventional in his religious views. He also reviewed the Origin in 1860, but his approach was to demolish, not explain, the arguments.

Had Mr. Darwin or his followers furnished a single fact to show that individuals change, in the course of time, in such a manner as to produce . . . species different from those known before, the state of the case might be different. But it stands recorded now as before, that the animals known to the ancients are still in existence, exhibiting to this day the characters they exhibited of old. The geological record, even with all its imperfections, exaggerated to distortion, tells now, what it has told from the beginning, that the supposed intermediate forms between the species of different geological periods are imaginary beings, called up merely in support of a fanciful theory. The origin of all the diversity among living beings remains a mystery as totally unexplained as if the book of Mr. Darwin had never been written, for no theory unsupported by fact, however plausible it may appear, can be admitted in science. . . . It would be superfluous to discuss in detail the arguments by which Mr. Darwin attempts to explain the diversity among animals. Suffice it to say, that he has lost sight of the most striking of the features, and the one which pervades the whole, namely that there runs throughout Nature unmistakable evidence of thought, corresponding to the mental operations of our own mind, and therefore intelligible to us as thinking beings, and unaccountable on any other basis than that they owe their existence to the workings of intelligence; and no theory that overlooks this element can be true to nature.

Both Agassiz and Gray came from conventional Christian backgrounds, but neither accepted the inerrancy of Genesis in explaining the origin and diversity of life. Agassiz, especially, had made geology one of his major research interests, and he was well aware that the fossil record showed different faunas in the different geological periods. This refuted the Christian belief that all species had been created at the same time and had remained unchanged to the present day. He was equally aware, however, of the absence of any fossil evidence that one kind of animal evolves into another. Fossils that were intermediate between major groups of organisms, as required by Darwin's theory, were nowhere to be found. Darwin himself was fully aware of these difficulties and made no effort to conceal the fact that they spoke against his hypothesis. For Agassiz, Darwin's theory failed to explain the variety of living things. His only suggestion, which had a powerful appeal for most individuals, was that the entirety of nature was due to the workings of intelligence. Thus he allied himself with Paley and the natural theologians.

The drama unfolded with even greater intensity in England. For most nonscientists the arguments offered in support of evolution often seemed irrelevant and difficult to understand: Mr. Darwin had entitled his book On the Origin of Species, but he admitted there was no direct evidence for even one species changing into another! Most scientists, including accomplished field naturalists, were equally unconvinced. The conventional view of the fixity of species accounted for the data quite satisfactorily, and the social pressures to believe that nature was created by Divine Will were hard to ignore. Consequently, powerful voices in science buttressed the popular reaction by proclaiming that Darwin's evidence would not survive careful scrutiny. Of the several reviews by important scientists that were published in the spring of 1860, most were negative.

In the many contentious debates about evolution that followed, many scientists along with most religious leaders and ordinary citizens stood on one side, while on the other side stood Darwin and a few stalwart supporters who believed that the concept of evolution provided a rational explanation for innumerable biological and geological facts and was worthy of serious study. The position of the supporters did not signify complete agreement that evolution by natural selection was true but only that it was a useful hypothesis to be tested.

The first notable public confrontation came on June 30, 1860—seven months after the *Origin's publication*—at a meeting of the British Association for the Advancement of Science at Oxford University. The event must have been eagerly anticipated, since the original room scheduled for the debate proved far too small, and the speakers and audience were moved to a larger chamber. No account of what transpired was published, but the story that has come down to us, in exaggerated form no doubt, is as follows: The mood of the audience—scientists, members of the clergy, and laypersons—was distinctly anti-Darwin. Because health problems prevented Darwin from attending, it remained for Thomas Henry Huxley (later dubbed "Darwin's bulldog") to support Darwin's position. The principal speaker to critique Darwin's views was the bishop of Oxford, Samuel Wilberforce. He was a prominent clergyman with a silver tongue, which had earned for him the sobriquet "*Soapy Sam.*" The fact that a prominent bishop rather than a scientist was chosen to refute Darwin's challenge shows that evolution threatened not just established scientific theories but the religious establishment as well.

Wilberforce knew very little about science, but the distinguished anatomist Richard Owen had coached him on the scientific arguments showing the improbability of evolution. The bishop did not fully understand Owen's arguments, it seems, and in any case he did not use them. But he did understand what evolution implied about human origins: that humankind was descended from ancient apelike creatures. Wilberforce's remarks to the very friendly audience were amusing and glib, and in his conclusion he turned to Huxley, who was sitting on the speakers' platform, and "begged to know, was it through his grandfather or his grandmother that he claimed his descent from a monkey?" The audience went wild. Asked by the chairman to respond, Huxley briefly outlined Darwin's views and then came in for the kill. After stating that he would not be ashamed to have a monkey for an ancestor, he turned to Wilberforce and added that he would, however, be ashamed to be connected with a man who used his great gifts to obscure the truth—implying that it was better to have descended from a monkey than from Soapy Sam. Huxley's remarks changed the mood of the audience dramatically.

Although nothing was solved by this superficial debate, many scientists realized that Darwin's position was novel and important enough to be considered carefully. T. H. Huxley became a dominant force in trying to help both scientists and ordinary citizens understand evolution. He was a gifted speaker and a first-rate biologist steeped in comparative anatomy, embryology, and general natural history. In 1860 Huxley gave a series of lectures, mainly related to evolution and other aspects of contemporary biology, to workingmen in London. These lectures were later published and reached an even larger audience. In 1863 he published Evidence as to Man's Place in Nature, in which he showed the close anatomical similarity of human beings with the great apes and suggested that this was sufficient evidence for all to be placed in the same family in the scheme of classification. Although Huxley was careful in drawing conclusions, a reader familiar with

Darwin's work would have suspected that Huxley thought the resemblances of the great apes and human beings were a consequence of their inheritance from a common ancestor. As time passed, Huxley became less sure that natural selection was the driving force behind evolution, as Darwin had proposed, and he was far from being alone among scientists in holding such doubts.

More professional debates followed within the scientific community, where harsh and demanding critics evaluated Darwin's data. Scientists were no less troubled than laypeople by the prospect of replacing God with a natural theory, and many went to great lengths to reconcile Darwin's views with the Judeo-Christian tradition. When Asa Gray reviewed the Origin in 1860, he suggested that Darwin's view of the relation of religion and the natural world was similar to that of the English philosopher William Whewell (1794–1866). Gray's reason for thinking so was that Darwin had placed a quote from Whewell opposite the title page of the first edition of the Origin: "But with regard to the material world, we can at least go so far as this—we can perceive that events are brought about not by insulated interpositions of Divine power, exerted in each particular case, but by the establishment of general laws." Gray (1860) suggested in his review what Darwin might have had in mind:

> We judge it probable that our author [Darwin] regards the whole system of nature as one which has received at its first formation the impress of the will of its Author [God], foreseeing the varied yet necessary laws of its action throughout the whole of its existence, ordaining when and how each particular of the stupendous plan should be realized in effect, and—with Him to whom to will is to do—in ordaining doing it. Whether profoundly philosophical or not, a view maintained by eminent philosophical physicists and theologians, such as Babbage on the one hand and Jowett on the other, will hardly be denounced as atheism.

Gray's position, following Whewell, was that God created the universe together with the rules that govern the interactions of matter and energy. At the end of creation He went away, leaving events in the natural world to spin out in agreement with His laws. These are the physical laws of mechanics, astronomy, and chemistry that scientists had been discovering since the Scientific Revolution, and evolution through natural selection might also belong on this list. Such a deistic worldview could easily accommodate both science and religion. *God is not eliminated, yet nature can still be studied systematically*—indeed, scientifically—because God's laws ensure that a given cause acting under defined conditions will always produce the same result. A scientist who seeks to discover the laws of nature will find them, whether he assumes they are ordained by *God* or are merely the ways that matter and energy naturally behave.

This reconciliation of creationism with science, in which *God* creates matter, energy, and their governing laws and then retires from the scene—a theological *Big Bang*, so to speak—was of little comfort to those laypeople who needed to believe in a personal *God* who was deeply concerned with their daily welfare. And as for scientists, there were other philosophical problems with this deistic worldview that had to be acknowledged. One of these had to do with the principle known as *Occam's razor*. William of Occam, a highly regarded English monk and philosopher of the fourteenth century, is best remembered for his philosophical position that "entities must not be unnecessarily multiplied." Among scientists, this meant that a minimum number of elements should be used in explaining a given phenomenon. Occam's razor would suggest that in explaining the diversity of life, there is no reason to invoke God to account for the "*laws*" of natural selection and variability. One could argue that these phenomena were not laws but were rather the inevitable consequences of life itself. This is not to say that God does not exist but only that He is an unnecessary part of the hypothesis.

To this day, people in science and in the Church have continued to wrestle with these problems and to try to adjust the new findings in evolutionary biology to traditional Judeo-Christian beliefs. And the struggle often takes the same forms that it did in the nineteenth century. Among people who believe in *God* are the strict fundamentalists, who simply deny the data provided by biologists and geologists and persist in believing in a *God* who created the world in six days and continues to intervene to guide the course of earthly events. A second group of believers might be called separationists; they assign to science the role of explaining the phenomena of nature and to religion the role of providing moral guidance, spiritual expression, and purpose. This position accepts that religion and science deal with different domains and are not in conflict. A third group, the modern deists, believe that *God* created the world and the laws pertaining to the interactions of matter and energy and since then has let the system run its course without further intervention.

Among those who do not accept any *God* are two principal groups. The agnostics (T. H. Huxley is credited with introducing this term) maintain that the existence of a *God* is unknown and unknowable. The atheists, however, deny altogether the existence of a *God*. Some scientists in this second category are quite vocal in their beliefs, but they should know better: one can no more prove that there is no God than prove that there is one—or more.

Darwin closed the Origin with this insightful and moving passage:

It is interesting to contemplate an entangled bank, clothed with many plants of many kinds, with birds singing on the bushes, with various insects flitting about, and with worms crawling through the damp earth, and to reflect that these elaborately constructed forms, so different from each other, and dependent on each other in so complex a manner, have all been produced by laws acting around us. These laws, taken in the largest sense, being Growth with Reproduction; Inheritance . . . Variability . . . [and] a Ratio of Increase so high as to lead to a Struggle for Life, and as a consequence to Natural Selection, entailing Divergence of Character and the Extinction of less-improved forms. Thus, from the war of nature, from famine and death, the most exalted object which we are capable of conceiving, namely, the production of the higher animals, directly follows. There is grandeur in this view of life, with its several powers, having been originally breathed into a few forms or into one; and that, whilst this planet has gone cycling on according to the fixed laws of gravity, from so simple a beginning endless forms most beautiful and most wonderful have been, and are being, evolved.

Possibly the most important point here is that these wonders "have all been produced by laws acting around us." Darwin was explaining the diversity of life not as a consequence of supernatural forces but as solely due to the interactions of natural things and processes.

What the Theory of Evolution Explains

The probability that a theory is correct becomes greater as it explains more and more data. This was the argument Darwin made. Despite his book's title, he did not provide detailed evidence in the Origin for the origin of any species. But he did assemble a vast quantity of data about variations of animals and plants in nature and under domestication, the struggle for existence, natural selection, laws of variation, instincts, hybrids, the paleontological record, geographical distribution, anatomy, classification, and embryology, and showed that numerous puzzles, otherwise inexplicable in natural terms, made sense in the light of evolution.

Fossil Record

The revolution in geology that Lyell had started several decades before Darwin was the recognition that the strata of sedimentary rocks are arranged in a sequence from oldest to more recent. The absolute ages of the individual strata could only be surmised, but relative age—older or younger—could be established beyond a reasonable doubt. Each major group of strata was found to have a unique population of fossil species. Some of the species might have closely similar counterparts in the strata above or below, and a few cases were known of sequences of slightly different organisms in successive strata.

Both of these phenomena—unique groups of organisms restricted to a single stratum and the apparent slow changes in a single type of organism in successive strata—were confirmed again and again as the nineteenth century progressed. While *Genesis* had no ready explanation for these data from the rocks, the findings could be easily explained by evolution. In fact, evolution required such fossil data. It also required that the time interval from the oldest known rocks to the youngest be immense. The extraordinary differences in the organisms in the oldest strata compared with the youngest strata could not have occurred in just a few millennia. Unknown millions of years were necessary. Scientists did not know then, as we do now, how long it had taken for the strata to form; but their total thickness was known to be many miles, and that amount of deposition would not be possible if creation had taken place only a few thousand years ago.

This history of life through time as revealed by the paleontological record was not absolute evidence for evolution, but evolution provided a satisfactory explanation for it. While the fossil evidence did not actually show the process of change of one species into another--it could not, since fossils are not living and so do not mutate, reproduce, and undergo selection—it did show closely similar fossils occurring in adjacent strata that could be explained best by invoking Darwinian evolution.

Linnaean Hierarchy and the Scala Natura

Many other scientific puzzles concerning anatomy, embryology, classification, and microscopic structure had no satisfactory answers in the absence of the concept of evolution. One of the broadest questions was why species seemed to fall so naturally into Linnaeus's hierarchial groups. Consider, for example, the major group to which human beings belong—the phylum Chordata. This is a heterogeneous group of creatures that includes some marine species—amphioxus and the tunicates—that can look like gelatinous blobs. Most chordates, however, are vertebrates—so-called because they have a series of bones on the dorsal side of the body that form the vertebral column. The major groups of vertebrates are fish, amphibians, reptiles, birds, and mammals. Vertebrates resemble one another in many ways other than just having a vertebral column. For instance, most of their organ systems are similar. The digestive, respiratory, excretory, nervous, skeletal, and reproductive systems are much the same in a trout, a frog, a lizard, a sparrow, and a white rat.

All vertebrates are constructed on the same general plan, that is, they are variations on the vertebrate theme. This astonishing observation could be explained as the consequence of divine creation: the Creator made all the vertebrates as variations on one basic theme. But the same data could also be explained by evolution: one would postulate that a relatively simple vertebrate—a fishlike ancestor—that lived a very long time ago was the progenitor of all vertebrates we see today around us and in the fossil record. Some descendants (modern fishes, whales) became adapted for life in water, others (birds, bats) for a life partially in the air, and still others for all the diverse terrestrial habitats. Nevertheless, all possess a similar basic structural organization of the body because they all descended from a common ancestor.

Figure is a highly schematic representation of evolution as a naturalistic explanation of the origin and diversification of the major groups of chordates. The origin of the chordates is represented at bottom left; the first organisms with the three main characteristics of this phylum, notochord, gill pouches, and a dorsal nerve tube, are thought to have evolved from an invertebrate ancestor. The simplest living chordates, the tunicates and amphioxus, are accepted as the closest living representatives of the most ancient chordates. The seven classes of vertebrates—chordates with a vertebral column—branched off at later times to give rise to many new groups, some of which flourished for long periods and then became extinct while others evolved into the vertebrates still alive. The approximate numbers of recognized living species in each class and examples of each are shown in parentheses. The very approximate dates for the beginnings of each class can be estimated by the numbers, in million years ago (mya), on the vertical axis.

Darwinian evolution also offers a plausible explanation for the scala natura. For example, the sequence of vertebrates—fish, amphibians, reptiles, birds, and mammals—observed in the fossil record matches

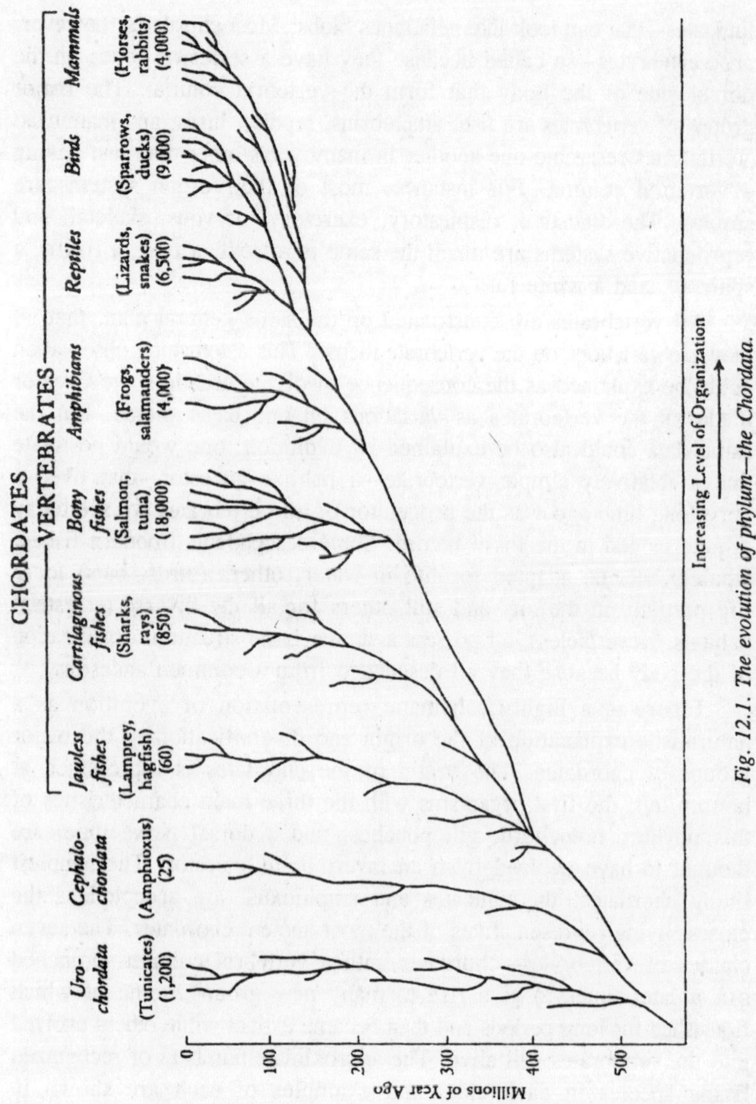

Fig. 12.1. The evolution of phylum—the Chordata.

the sequence of increasing complexity of anatomical organization: fish, it turns out, have the least complex bodies, while birds and mammals are the most complex. These parallel results suggest strongly that the underlying cause of the sequence of increasing complexity might be evolution. Both the paleontological data and the anatomical data make sense if we assume that some fishes evolved into amphibians, some amphibians evolved into reptiles, and some reptiles evolved into birds, while other reptiles evolved into mammals. Because the fossil record at the time Darwin was writing was still sparse, naturalists would have thought only that the data were consistent with evolution but not absolute proof. Later fossil discoveries would be more and more convincing.

Cellular Makeup of Both Plants and Animals

Another extraordinary observation that evolution explains much better than Genesis does is that the bodies of all animals and plants are composed of the same basic structural unit, the cell. Evolution explains this surprising evidence of relatedness between plants and animals by hypothesizing that the two kingdoms share a common, remote ancestor composed of cells, or perhaps comprising just one cell, like the bacteria and protozoans still living today. In the last part of the nineteenth century it was next to impossible to find fossil evidence for cells, which are microscopic in size and lack hard structures such as bones, teeth, and scales that fossilize readily. But later discoveries have supplied the missing data.

Embryonic Development

The theory of evolution has also proved a powerful tool for explaining some otherwise puzzling aspects of embryonic development. One of the most dramatic and complicated examples concerns the ear bones of all vertebrates except the fishes. The tympanic membrance and the ear bones conduct sound waves from outside the head to the inner portion of the ear, enabling hearing. Fish have neither a tympanic membrane nor ear bones. Amphibians, reptiles, and birds have a tympanic membrane and a single ear bone, the stapes, in each ear. Mammals have a tympanic membrane and three ear bones, the malleus, the incus, and the stapes. If indeed mammals evolved from reptiles, what could be the origin of those two additional bones, the malleus and the incus?

The first part of the answer was provided by the German embryologist Karl Reichert in 1837. Reichert found that although the adult mammal has a single bone, the dentary, in the lower jaw, in the embryo there is another bone, the articular. The upper jaw also has an extra bone, the quadrate, which together with the articular in the lower jaw forms the embryonic jaw joint. This condition in the embryo is the adult condition in reptiles, birds, and amphibians. Reichert observed that in the course of development the quadrate and articular of the mammalian embryo detach and move to become the malleus

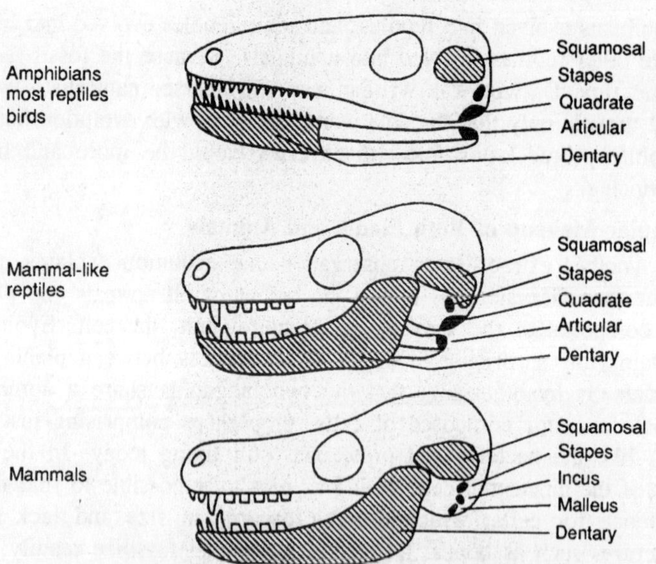

Fig. 12.2. The development of the mammalian jaw and ear bones.

and incus of the adult ear. It was suggested, therefore, that in the course of evolution, the quadrate and the articular that formed the articulation of the jaws in other vertebrates had evolved into the incus and malleus of the mammals. Yet how could a switch from one type of joint to another occur? Certainly it could not be instantaneous—evolution is too slow for that. Then could one imagine an intervening form that had two articulations? Could a jaw with two joints really function? Was it possible to obtain evidence to support this hypothesis?

Surprisingly, the answer to these questions proved to yes. Long after embryologists had proposed the transformation of the reptilian jaw bones into the jaws and ear ossicles of mammals, a series of fossils from South Africa and North America revealed that the hypothesis was correct. The fossils belonged to a group of rather early reptiles known as the mammal-like reptiles. And they formed a sequence that showed the conversion of the quadrate and articular into the incus and the malleus, and the transition from the reptilian to the mammalian type of jaw articulation. Once again, the data demanded by theory became available.

Another embryonic puzzle that evolution resolves is the development of the vertebrate kidney. Figure is a schematic representation of the development of the kidneys, which start as tubules linked by a duct that carries urine to the posterior part of the body,

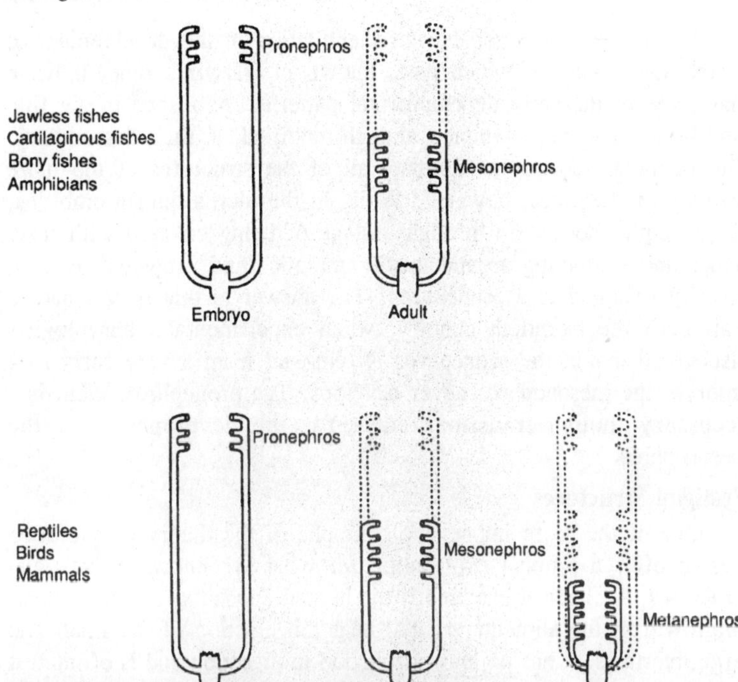

Fig. 12.3. Developmental changes in the vertebrate kidney.

where it is expelled. The first kidney to form in the embryos of all vertebrates is the *pronephros* (derived from the Greek pro, meaning "in front," plus nephros, meaning "kidney"). The kidney of the adult hagfish, the most primitive living jawless fish known, is thought by some anatomists to be a persisting portion of the embryonic pronephros. The older embryos of all other vertebrates next develop a kidney further back in the body, the *mesonephros* (mes meaning "middle"). The mesonephros remains the kidney of the adult in the lower vertebrates—the cartilaginous, and bony fishes and the amphibians. The embryos of the higher vertebrates develop first a pronephros and then a mesonephros, but as older embryos form a third kidney, the *metanephros* (meta meaning "posterior") which remains as the kidney in the adult. Based on these observations, the hypothesis is that the first vertebrates probably had only a pronephros and that the higher fishes and amphibians evolved a *mesonephros* and finally the reptiles, birds, and mammals evolved a *metanephros*. The accepted scientific explanation is grounded in the concept of evolution: the more advanced vertebrates recapitulate the stages of kidney development in the lower vertebrates.

The numerous examples of recapitulation in the development of vertebrates, many of which were known in Darwin's time, indicate that some of the early developmental patterns established in the first vertebrates have been retained, though modified, in their descendants. The mammals do not recapitulate all of the structures of the more primitive vertebrates, however. While in the uterus human embryos, for example, do not go through a stage of being covered with fishy scales and swimming around in the amniotic fluid propelled by fins. So why recapitulate a pronephros? The answer to that puzzle had to wait until the twentieth century, when experimental embryologists discovered that if the pronephros is removed from a very early frog embryo, the mesonephros never develops. The pronephros, then, is a necessary though transitory stage for the development of the mesonephros.

Vestigial Structures

One of the more impressive triumphs of the theory of evolution was to offer a rational explanation for what are known as vestigial structures. A classical example is the human appendix, a small tubular outgrowth of the alimentary canal near the junction of the small and large intestines. It has no known function in digestion and is of interest only when it becomes infected and has to be removed surgically. A study of other mammals has shown, however, that the appendix is an important structure in some species. The rabbit, for example, has a very long and well-developed one in which bacteria live that digest complex carbohydrates, such as cellulose, which make up a large component of the rabbit's food. Human beings do not have these cellulose-digesting bacteria and thus do not need an elongated appendix to serve as their home, so natural selection has never promoted its development.

There are numerous examples of vestigial structures throughout the animal and plant kingdoms. The ancestors of snakes walked on four legs. Most snakes today have no vestiges of legs, but a few do—tiny bones under the skin where the hind legs should be. Some whales, whose ancestors lived and walked on land, have tiny bones inside the body where the hind legs should be. The whales, however, have retained their front limbs as flippers that aid in locomotion and in maintaining stability. The tail is their main organ of locomotion.

There seem to be few biological puzzles for which Darwinism does not suggest a plausible answer. In some dramatic instances confirmation from the fossil record is possible. And there is nothing in the fossil record that falsifies the theory of evolution. At the close of the nineteenth century, instead of the age-old religious explanations that were still widely held, Darwin offered a useful theory based solely on logical and natural principles: evolution through the mechanism of natural selection acting on variation in a finite environment, resulting in descent with change. Yet evolution is a historical science, which means that very little can be verified by direct observation. The data for ancient organisms, if they persist, lie buried in the geological strata. We cannot replay the tape of the Earth's history and watch dinosaurs evolving into birds. We can do no more than search for indirect data that provide evidence for such a transition.

13

CREATIVE REPRODUCTION

On the face of it, the reproducibility of organisms, their similarity from one generation to the next, would seem to imply that some sort of plan or set of instructions is being followed as they form. Internal painting, the elaboration of hidden colours, occurs in pretty much the same way every time a fruit fly develops, suggesting that it is much more like slavish manufacture than a creative process. To pursue this issue any further we will need to take a closer look at what underlies reproducibility. What determines the consistent development of organisms every generation? Once this has been explored, we will be able to make a more informed comparison with art and see whether there is a fundamental contradiction between creativity and reproducibility.

Although biological reproduction tends to produce similar individuals each generation, they are not all exactly alike. We are most aware of this in our own species, as variation in the facial features and physique from one person to the next. But we might detect a similar level of variation in other species if we were only as sensitive to their individual characteristics. The ornithologist Peter Scott, for example, learned to recognize individual swans by noting the colour markings on their bills. It is just that normally we are less aware of variation in other species because it does not seem terribly important to us. Distinguishing between people is an essential requirement of everyday life but we can get by without being able to recognize individual daisies or fruit flies.

Much of the variation between individuals in a species comes from genetic differences—slight variations in their sequence of DNA. If the DNA of two individuals is exactly the same, as in *identical twins*, they resemble each other much more closely than individuals with distinguishable DNA sequences. The extent of biological reproducibility therefore depends on the similarity in genetic make-up. For maximum consistency in development, the DNA sequence in the developing egg should always be the same: the fertilized egg should always begin with exactly the same genes. When we say that development is a completely reproducible process, it is in the context of a very similar, preferably identical, genetic background.

In addition to genetic make-up, the environment also influences the way an organism develops. A seed sown in one type of soil and climate may produce a very different looking plant from a similar seed grown under other conditions.

In some cases, you might even take two plants as belonging to different species simply because they have grown under different conditions. Similarly, environmental conditions have important effects on the way animals develop. If fruit fly larvae are exposed to a relatively high temperature for short times, they will often develop into adults showing major abnormalities in their body structure; defects that are just as severe as those produced by mutations in genes. Another striking effect of the environment is illustrated by certain species of butterfly that develop in different colour forms from one season to the next, say one form appearing in spring, the other in summer (a phenomenon called polyphenism). The two colour forms may look as distinct as two different species, yet they only differ because of the environmental circumstances that prevailed at the time of their development (the relevant environmental factors can include temperature and day-length). In the case of mammals, the environment of the embryo is carefully controlled in the mother's womb and buffered from the conditions outside the mother. But if the embryo's environment is altered in some way, it may lead to alterations or abnormalities in the way the embryo develops. For example, if a mother has German measles (rubella) during early pregnancy, the resulting child may be born with severe abnormalities. The *rubella virus* is able to cross the placental barrier between mother and child, and infect the embryo, modifying its development.

When we say that an organism develops in a consistent and reproducible manner, it is always in the context of a specified environment and genetic makeup. Both of these factors must be defined from the outset if development is to proceed in a predictable way. To make a fair comparison between development and artistic endeavour, we should therefore standardize the initial conditions for painting a picture in the same way that we have done for a developing organism.

We can think of the surroundings of an artist as equivalent to the environment of a developing organism. In the case of Leonardo painting the Mona Lisa, for example, there is the subject herself and the background around her. The general lighting conditions are also a very important ingredient, as Leonardo himself was at pains to point out: when you wish to paint a portrait, paint it in bad weather, at the fall of the evening, placing the sitter with his back to one of the walls of the courtyard. Notice in the streets at the fall of the evening when it is bad weather the faces of the men and women—what grace and softness they display! Therefore, O painter, you should have a courtyard fitted up with the walls tinted in black and with the roof projecting forward a little beyond the wall; and the width of it should be ten braccia, and the length twenty braccia, and the height ten braccia; and you should cover it over with the awning when the sun is on it, or else you should make your portrait at the hour of the fall of evening when it is cloudy or misty, for the light then is perfect.

As part of the general environment we might also include the musicians and jesters that, according to the biographer Vasari, were employed by Leonardo so as to keep the subject of the Mona Lisa smiling while he painted her. To standardize the environment of a painting, we should therefore only look at works that have been produced under the same conditions, keeping the subject, light and all the other elements in the surroundings fixed.

These conditions alone do not, however, determine the type of painting that is produced. Much depends on the particular artist involved. Compare the *Mona Lisa* to the portrait of an Italian woman, L'Italiana by Vincent van Gogh. Clearly they differ greatly in style even though the subjects are comparable. If you were to imagine transporting exactly the same model that Leonardo was painting, together with her surrounding courtyard and entertainers, to the late nineteenth century for van Gogh to paint, there would be little difficulty in distinguishing the result from the *Mona Lisa*, even though the immediate environment would have been the same in each case. This is because each painter works in a very different way. It is not simply that Leonardo's painting is more realistic than van Gogh's. Leonardo used various stylistic devices to portray the subject in a particular way. The shadows, particularly around the eyes and mouth of the *Mona Lisa*, are left purposely diffuse and blurred to create an ambiguous and slightly mysterious effect (a technique invented by Leonardo, called sfumato).

Also the background that Leonardo paints is an imaginary landscape, not the one that would have been behind her. Conversely, although van Gogh's style with its strong brush strokes may seem less realistic, the painting may convey certain features of the sitter's personality and appearance more effectively. Leonardo and van Gogh simply painted in different styles, each of which can be seen to have its own merits.

We might compare the paintings of Leonardo and van Gogh to the development of two different species. Even if the immediate environment were to be kept similar, each would produce a very different result, In the case of the painters, this is because of their distinct cultural and biological heritage. Leonardo was an illegitimate child of the fifteenth century, brought up during the height of the Italian Renaissance. Vincent van Gogh had a different biological heritage and was born four hundred years later when Impressionism was coming to its peak. This means that when each one starts a painting, they come to it with a very different set of circumstances behind them. Their brains operate and react in different ways because of their distinct biological and cultural heritage.

Similarly, two species can develop in very different ways because they have a distinct evolutionary history. In this case, the legacy of the past is held in the genetic make-up and properties of the fertilized egg. If these differ substantially, so will the appearance of the organism that develops, even if the environment is similar.

The reproducibility of organisms with a given genetic make-up and environment should therefore be compared to a series of paintings in which both the artist and the surroundings are standardized. We might want to look at a series of paintings by Leonardo, with him always in the same surroundings. Strictly speaking, we really need to standardize not just the artist but his state of mind at the beginning of each painting. This would be the equivalent of identical twins starting with exactly the same genetic make-up. But now we come to a problem. After Leonardo has painted the *Mona Lisa* once, his state of mind will not be the same as when he started. The very act of painting the picture will have changed his outlook. For one thing, he will now have the painting of the *Mona Lisa* before him whereas this was not available when he first started out. If he were to paint the same subject a second time he would know of the previous result. Perhaps he would avoid the second painting being too much like the first, exploring novel features that were not covered so well in the first painting. Or he may no longer be interested in the subject so much after having already spent a lot of time on it. Whatever the case, he will not start the second painting with the same mental state as he did the first time round.

To get around this problem, we might imagine that Leonardo has a bout of amnesia after he finishes each painting. Suppose that every time he completes the *Mona Lisa*, he forgets everything he did previously during the painting. He wakes up in the same mental state after finishing it off as when he started to paint the picture the first time. Of course there also has to be a conspiracy amongst his friends so that someone hides the finished painting from him each time and no one tells him that he has done it all before.

Given these admittedly peculiar circumstances, Leonardo might start to produce a whole series of versions of the *Mona Lisa* over time. We now have a situation that is equivalent to the development of an organism with a fixed genetic make-up in a standardized environment. Of course the numerous *Mona Lisa* paintings might not look exactly the same because of chance variations that are out of our control, but then again identical twins do not look exactly the same.

The point is not so much to do with whether the paintings are precisely the same each time, but with the nature of the painting process. An outsider who is unaware of the conspiracy, and simply sees all the versions of the *Mona Lisa* being churned out, might conclude that Leonardo is simply manufacturing lots of slightly different copies of the same painting. Based on the consistency of the output, it may seem they are being produced by copying from an original or by always following the same routine set of instructions.

But this is not at all the way that Leonardo would see it. To him each painting is a genuine and novel creation. He is not copying because he does not know what the finished picture will look like. Nor is he manufacturing them according to an independent plan or set of instructions. He is creating the pictures through a highly interactive process in which there need be no dear separation between plan and execution. Each time he might start out with some idea of the sort of picture he wants to paint and apply a few brush strokes. He may react to these in one way or another, adding some here, modifying some there. The picture emerges through a creative process rather than one of fabrication or copying. Although all the versions may look similar, they have been created as originals, without any advance knowledge of the final outcome. The external observer who judges that Leonardo is copying or manufacturing based on the consistency of the output would have completely misunderstood what Leonardo was actually up to.

At this point the reader might object that although Leornardo need not have had a detailed plan in mind while creating the *Mona Lisa*, he nevertheless might have had one. Perhaps Leonardo did have a very precise idea of how he was going to paint the picture: a dear mental vision of how he was going to proceed before he even put his brush to the canvas. Wouldn't this mean that he was being creative while at the same time following a plan? In practice most artists do anticipate the outcome of their work in detail, because this would be to deny the importance of how they interact with the medium. Nevertheless, let us ignore this and suppose, for the sake of argument, that Leonardo did have a precise plan for all the brush strokes needed to paint the Mona Lisa, and a dear vision in mind of what the final picture would look like before he painted it. The process of painting would then involve a clear separation between plan and execution.

But this would raise the question of where the plan in Leonardo's mind itself came from. How did Leonardo come up with this detailed vision? It is no good saying that the vision was itself derived from an earlier mental plan, because this would simply beg the question of where this earlier plan came from, leading to an infinite regress. The answer would surely be that the vision was a creative product of Leonardo's mind: the plan was itself the outcome of a highly interactive mental process. This process may have occurred entirely within Leonardo's brain, or perhaps it might have also involved some preliminary sketches, carried out by Leonardo to help him formulate the plan more clearly. Either way, the plan would have arisen through an interactive process that could not itself have been a matter of following a plan (otherwise we are back to an infinite regress). Now the key point here is that if Leonardo had set about painting the *Mona Lisa* in this way, the most creative part from his point of view would have been coming up with the plan or vision, rather than the act of painting. Once he had arrived at a detailed mental conception of how to paint the picture, executing the plan by applying paint to the canvas would have been a purely mechanical exercise, equivalent to the process of writing down a poem once the lines have been thought up. To put it another way: if he had a perfect vision in mind of what the final picture would look like, the act of painting would simply correspond to copying this vision. So even in this case, the creative process would not involve the following of a plan. Rather, it would be concerned with the events before the painting had begun, when Leonardo was coming up with his detailed vision.

If the *Mona Lisa* had arisen in this way, we could distinguish between two phases i its not production: a primary creative process of arriving at a plan, and a secondary technical process of executing the plan. So long as Leonardo wakes up from his bouts of amnesia in a mental state that precedes both of these phases, he would still feel he was being genuinely creative each time, although in this case the creativity would occur in the phase of coming up with a plan rather than during the application of the paint. He would still wake up with no clear idea of what the final painting would look like, having to arrive at his vision each time through a creative process. As before, the external observer seeing similar paintings being churned out might misunderstand what was going on. It might look like a case of routine manufacture; whereas it actually involves a highly creative process that is not primarily derived from following a plan.

We can imagine a situation in which a process is both genuinely creative and yet has a more or less reproducible outcome. Of course, in practice artists do not have bouts of amnesia every time a painting is finished, so it is a reasonable assumption that a series of very similar pictures are more likely to have been copied or fabricated in some way than created. Now when it comes to development, we are struck by the reproducibility of organisms and are therefore drawn to analogies with a process of manufacture according to a plan. We naturally equate development with someone consistently following instructions. But in my view this is misleading. We are making the same mistake as the external observer seeing the output of Leonardo the amnesiac. The process of development is much more like creating an original than manufacturing copies, in the sense that the final product, the adult, is not there from the beginning but gradually emerges through a highly interactive process, in which each step builds on and reacts to what went before in a historically informed manner. Like the amnesiac Leonardo, the fertilized egg has no detailed sense of what the final picture will look like. It is not so much that the egg forgets how the previous picture looked; it never got to see the picture in the first place. It is simply a tiny part of the earlier canvas, one cell amongst many others that nevertheless carries all the information needed to produce another canvas, given certain environmental conditions.

The notion of each step being historically informed is very important here, and distinguishes development and human creativity from many other sorts of interactive process. There are many processes in which the various steps are not historically informed, even though the steps build on each other in complex ways. Take the formation of a snowflake.

This depends on a complex sequence of events in which water solidifies at the surface of a growing ice crystal. This process is highly interactive: the shape of the crystal influences how it will grow, which in turn will influence the subsequent shape of the crystal. Each step in the growth of a snowflake therefore depends on what went before and influences what is to come (slight variations in the growth conditions may therefore affect the final form in quite complex ways, accounting for the many different shapes of snowflakes). However, none of these steps is historically informed: the way a crystal grows is not influenced by the experience of crystals that have grown in the past. It simply depends on the prevailing physical conditions. By contrast, each step in biological development is informed by the evolutionary history of the organism, involving countless previous generations of natural selection. The creativity of human beings also depends on their biological inheritance, as well as on their cultural heritage and experiences. In this respect, the development of a fertilized egg and a creative process are much more similar to each other than to processes in which there is no historical input of this kind.

The fertilized egg is driven by some mysterious creative force. The egg is actually thinking about what it is doing as it develops. My point is that human creativity can give us a better overall sense of the nature of developmental processes than comparisons with fabrication according to a plan. If we think of development in terms of manufacture, we will start to look for a set of instructions as distinct to whatever it is that follows them, or we will search for an independent plan that is then executed, or try to identify software as distinct from hardware. It is the notion of reproducible manufacture that impels us to look for these distinctions. B as we have already seen with theut exercise of internal painting in the previous chapter, and as will become more apparent during this book, development is an interactive process in which each step builds upon and reacts to what went before in a historically informed manner. We will try in vain to make a separation between plan and execution because the two are deeply interwoven. And to my mind this is similar to what happens in creative processes. Being external observers, the consistent output of development leads us to equate it with manufacture. But once we delve into the internal mechanisms, we see they have more features in common with creative activities, even though they give reproducible results.

14

ORIGINS OF GENESIS

The Old Testament of the Christian tradition consists of the sacred scriptures of the Jewish people. The antecedents of *Genesis*, the first book of the Hebrew scriptures, date to the very dawn of recorded history and possibly to the early days of civilization in the ancient Near East, where the earliest cities seem to date to the few centuries before 3000 b.c. The Hebrews were the Semitic group from which both Judaism and Christianity arose. The Semitic people first entered history as tribes of nomads grazing their herds in the grasslands of Mesopotamia (now Iraq) and the Arabian peninsula. Five major groups are recognized: Akkadians, Canaanites, Phoenicians, Hebrews, and Arabs. The Akkadians settled in Mesopotamia; the Canaanites, Phoenicians, plus the Hebrews eventually migrated west to the eastern shore of the Mediterranean. The Arabs continued to live in Arabia to the south. Much later, after the death of their prophet, Muhammad, in 632 a.d., the Arabs swept throughout the Mediterranean world as conquerers. They developed the Islamic religion with its holy book the Koran. Eventually most of the Semitic people abandoned the nomadic life of their ancestors and adopted agriculture, lived in cities, and developed complex political, social, and religious institutions.

The basic source for the early history of the Hebrew tribes is the Old Testament, plus a little information from archaeology and the records of neighbouring people. For example, ancient Egyptian records, which are fairly extensive, make vague references to people who may have been the Hebrews. According to Old Testament sources and a modicum of other information, at least some of the Hebrew tribes seem to have been in ancient Sumer in southern Mesopotamia—possibly the site of the first civilization. About 1900 b.c. their leader was Abraham, according to ancient tradition. Later the tribes migrated westward to what is now Palestine and on to Egypt, possibly between 1700 and 1600 b.c. There, according to their traditions, they were enslaved. Sometime between 1300 to 1250 b.c. Moses led them back to Palestine, their Promised Land. Palestine was then known as Canaan, and it was occupied by the Canaanites, another group of Semites who were culturally more advanced than Moses' Hebrews, some even living in cities. Fertile land was scarce, and because the Canaanites were not inclined to abandon their territory to the newcomers, there was strife between the two groups for several decades. In fact, about all the Hebrews were able to conquer at first were the rather infertile hills to the east of the fertile lowlands that bounded the Mediterranean.

The period around 1200 b.c. was an exceptionally violent time in the eastern Mediterranean world for reasons not well understood. Some historians suggest that a mysterious "*Sea People*" traveled about widely, leaving destruction in their wake. The palaces of the Mycenaeans in what is now Greece were leveled; Troy fell; the Hittite Empire collapsed; and devastation extended from Greece through Turkey to the southern part of Palestine. Only Egypt survived relatively intact, having repulsed the Sea People in battles along the Nile delta. What had been an impressive Bronze Age, when the heroes described by Homer in the Iliad and the Odyssey lived, was replaced by a dark time that lasted several centuries.

One group of Sea People who entered Canaan during this period of turmoil were the Philistines. They proved to be far better warriors than the Canaanites or the Hebrews—a major reason being that their swords and spears were made of iron rather than bronze. The eastern Mediterranean world had embarked upon the *Iron Age*.

The Hebrew people, after losing some of the land they had conquered from the Canaanites, recognized the need for a more efficient political structure. Up to this time (about 1225–1025 b.c.) the Hebrew tribes had been loosely associated and led by charismatic religious and political leaders. Around 1025 b.c. the tribes began to consolidate, with Saul as their king. By the reign of David—next in line after Saul— the northern and southern tribes were united, and their armies successfully repulsed the Philistines. David and his son and successor, Solomon, were extravagant to the extreme and supported their projects with heavy taxation—a cause of considerable social unrest. So when Solomon died in 922 b.c., the united Hebrew kingdom split, with the ten northern tribes becoming the Kingdom of Israel and the two southern tribes, the Kingdom of Judah.

For the next eight centuries a series of invasions kept the Hebrews in turmoil. In 722 b.c. Assyria conquered the Kingdom of Israel and dispersed its inhabitants, who disappeared from history as organized communities. In 586 b.c., King Nebuchadnezzar of Babylon overwhelmed the Kingdom of Judah, destroyed Jerusalem, and took many prominent Jewish people back to Babylon—an episode known to history as the Babylonian captivity. Next came a period of Persian rule, from 539 to 332 b.c. Then Alexander the Great (356–323 b.c.) of Macedonia conquered most of the vast area from Greece to India and unified the Middle East both culturally and politically. After Alexander's death, Palestine was ruled by his generals, and much later it became part of the Roman Empire.

In 70 a.d. the Romans suppressed a Jewish rebellion and destroyed Jerusalem and the Temple. Many of the Hebrew people migrated to the various Roman provinces throughout the Mediterranean world—still another diaspora. What the Assyrians and Babylonians began, the Romans continued. Thus the Hebrew people were able to live free and united in their Promised Land for only relatively short periods—the few hundred years before 922 b.c. and less than a decade in the second century b.c. It is only in the twentieth century that the Jewish people have reestablished a nation of their own.

Scholars generally accept that the Hebrews began to develop their religious beliefs when they were seminomads. The first objects of their worship were probably the sun, the moon, springs, and mountains as well as mythological creatures and supernatural forces. In those early days, the gods or other objects of worship tended to be geographically restricted, with each cultural group worshipping its own gods. And so as they traveled the Hebrews would have encountered a bewildering variety of deities throughout the Middle East, from Atum and Horus in Egypt to Baal and El in Canaan.

The defining event in the Hebrews' religious development was the movement toward a *single god*, *YHWH* (or Yahweh when vowels are added). Moses is given considerable credit for this conversion from polytheism to monotheism. The name *Yahweh* has been variously translated as "He is," "He causes to be," or "I am who I am." *Yahweh* was in part a war god, but Jewish tradition credits Yahweh with transmitting to Moses many of the moral values and customs that influence our lives to this day—the Ten Commandments being foremost.

The emphasis on moral values continued with the great Hebrew prophets Isaiah, Hosea, Amos, Micah, and of course Jesus.

There is scant information about the development of the religion of the Hebrews before the first millennium b.c. For a long time the sacred traditions were probably transmitted solely by word of mouth. Homer's Iliad and the Odyssey are examples of other stories transmitted orally long before they were produced in written form. An oral tradition implies the presence of special persons, such as priests and elders, who were the custodians and transmitters of sacred beliefs and religious practices. Oral transmission always carries a high risk of error, because the transmitter might forget some portions of very long stories and might alter others.

The word bible is derived from byblos, meaning a roll of papyrus, which together with leather, clay tablets, and stone were the writing materials in the ancient Mediterranean world. Byblos was also the name of a Phoenician city known for the production and export of papyrus. Single sheets were prepared by arranging layers of reeds from the papyrus plant at right angles to one another and then pressing and drying them. The result was a very durable and easily handled writing material. These flexible papyrus sheets could be attached to one another to form writing surfaces of various lengths, and the documents could be rolled and unrolled repeatedly. Pens were made from sharpened reeds, and ink was produced by mixing carbon with oil. The *Bible*, then, began as a collection of papyrus scrolls containing the sacred writings of the *Hebrews*.

Copying sacred texts onto papyrus scrolls was less prone to error then oral transmission, but complete accuracy was impossible. In the seventh century a.d., printing from wooden blocks was practiced in China and Korea. But in the West paper did not become available until the Middle Ages, and movable type was not invented until about 1450. Only with these technological advances could accurate and relatively permanent multiple copies of important documents such as the *Bible* be made.

Many of the sources that became the books of the *Hebrew Bible*, or Christian Old Testament, were produced as papyrus scrolls in the last millennium b.c. Three sources are of critical importance for the first portion of Genesis, where the stories of creation are told. They are known among scholars by the abbreviations J, P, and R.

The J source is the oldest and is thought to have been first produced in the tenth century b.c.—possibly between 950 and 800. J stands for

Jahweh, the German spelling of *Yahweh* (many of the nineteenth-century scholars studying the origins of the Old Testament were German). This text was created shortly after the death of King Solomon and the separation of his realm into the kingdoms of Israel and Judah. P is for priestly, since it was the work of Hebrew priests who came later, possibly between 700 and 500 b.c. During those years the Hebrews were under the control of the Assyrians and the Babylonians. R is for redactor, referring to the editor or editors who in the fifth century, possibly between 450 and 400 b.c., combined J, P, and other sources to produce Genesis—probably in much the same form as we know it today. Two other major sources, E and D, are basic to the text of later books in the Old Testament but not to Genesis.

A word should be said about the first five books of the Bible: Genesis, Exodus, Leviticus, Numbers, and Deuteronomy. In the Hebrew tradition they are known also as the *Law*, the Law of Moses, the Pentateuch (meaning "five books"), or the Torah. They are of special significance to the Jewish people, as they are considered a direct communication from God through Moses to the Jews. In recent decades biblical scholars have questioned Moses' role in writing the Torah, in part because the last verses of Deuteronomy describe his death—and writing about one's own death is a difficult feat for any author to perform. Another work of major significance to Judaism is the collection of the opinions of rabbis relating mainly to moral issues and conduct. These opinions were orally transmitted until about 200 a.d., when they began to be preserved in written form as the Mishna; this was expanded to become the Talmud in the fifth century a.d.

The date when the Hebrew scriptures were first assembled to produce a single volume is unknown. Beginning in the sixth century a.d., successive generations of rabbis known as the Masoretes attempted to compare the surviving documents and rid them of errors and differences to ensure the consistency of the texts. This approach was quite different from that taken by the compiler of another sacred text, the Koran, which consists of the revelations of God to Muhammad as written down by his secretary. About 650–651 a.d. the Caliph Othman assembled the secretary's notes and then took the precaution of destroying all other notes and versions. Thus he did not have to deal with a confusing and contradictory set of documents, as did the redactor(s) for Genesis.

The oldest extant copy of the Masoretic text was transcribed in 950 a.d. by Aaron ben Asher. This was about two thousand years after the first scrolls of the earliest Genesis stories may have been prepared, and of course, they would have been copied numerous times in the interim. Nevertheless, there is good evidence that the surviving Masoretic text has a high level of accuracy: it is nearly identical with the *Dead Sea Scrolls*. These scrolls date from roughly 250 b.c. to 135 a.d., and were rediscovered from 1947 onward by Arabs in caves near the *Dead Sea*. A copy of the Masoretic text, dated 1008 a.d., is the basis of today's standard version of the Hebrew scriptures.

Far older is the Septuagint, a Greek translation produced in Alexandria, Egypt, beginning in the third century b.c. from Hebrew originals that no longer exist. The Septuagint was prepared for Jews living in Egypt who had lost their knowledge of Hebrew and needed a version of their sacred traditions in a language they could read. That language was Greek, the dominant language in the Near East after the conquests of Alexander, a great promoter of Greek culture and language. There is a wonderful story about the origin of the Septuagint, which means "70." The Egyptian king Ptolemy II (Philadelphus) ordered the translation, which was supposedly completed by 70 or 72 scholars working independently for 72 days. The purpose of isolating the scholars was to see if each of them was truly inspired by God: if they were, all the translations should be identical. At the end of the allotted time the 70 translations were examined and, lo and behold, they were exactly the same. That convinced the king and all others that the translations must have been the inspired, accurate *Word of God*.

An important source of the *Christian Bible* is the *Vulgate*. This is a Latin translation made by Saint Jerome around 400 a.d., based on numerous variant Hebrew texts that are no longer available and also on the Septuagint. Another of his sources was the Hexapla, which had been prepared by the Christian scholar and philosopher Origen (185–254 a.d.). It consisted of six parallel columns, each with a different version of the sacred texts in Hebrew or Greek. Not knowing which column represented the true Word, Jerome had to make choices and change the text when he suspected that errors had been made by copyists. The usefulness of his work was subsequently much reduced by careless copying and other modifications. A somewhat improved Paris text of the Vulgate was printed in 1450–1452 by Gutenberg. It, too, changed over the centuries, but it remains the standard Bible of the Catholic Church.

Thus the Bibles of today have had a long and varied history. Generation after generation of scribes copied the sacred documents, introduced errors, corrected presumed mistakes, added new material, and made translations. The advent of printing reduced the variation, but it also meant that a few variants were arbitrarily selected as closer to the original meaning, while other versions were ignored. Today there are only a handful of standard translations in common use, and there are still significant differences among them.

The desire, of course, has always been to come as close as possible to what was originally said in the long-lost original Hebrew scrolls. This would not be an easy task even if an original scroll were available. Ancient Hebrew had almost become a ritual language before the time of Christ, and Aramaic was becoming the common language of the Jews. Moreover, Hebrew was originally written only with consonants, which could lead to serious problems in knowing what word was meant. Consider the problem of knowing the meaning of an English word written lk. It could be like, lake, look, leek, or luke, depending on which vowels are used. Macintosh (1972) provides this more complex example (slightly modified): "the cat was on the mat" would be written "*thctwsnthmt*" if the words were run together, as was the practice, and the vowels left out. Depending on the vowels inserted, the sentence could be read as "the coat was on the mat," "the cat was in the moat," or "the cut was in the meat." It was only about a thousand years ago that dots and dashes were added to the consonants of written Hebrew to indicate vowel sounds.

The problem of no vowels is actually not as serious as it might seem. Even when the text of *Genesis* had been written on scrolls, the oral tradition continued and the scriptures were read aloud. Thus the words written only with consonants were understood since they were also heard. So long as the spoken words were correct and the documents had not suffered through poor copying, accuracy would continue.

Nevertheless, the problem of knowing the meaning of many ancient Hebrew words continues to this day. Does an unintelligible word represent a copyist's error or is it a word for which the meaning has been lost? Scholars can sometimes resolve the problem by reference to other Semitic languages. For example, a word thought to mean only "*to know*" in Hebrew means both "*to know*" and "*to be tamed*" in Arabic. This suggests that Judges 16:9, which refers to Samson, could be translated "And his strength was not tamed" instead of "So his strength was not known." It should come as no surprise, therefore, that there are differences—often very important differences—between versions. For example, Isaiah 7:14 in the Septuagint reads "The virgin will conceive and bear a son," whereas in the Masoretic text virgin is replaced by young woman.

There are problems of translation even with the very first sentence in Genesis. The first phrase in Hebrew is bereshith bara elohim. Traditionally this has been translated "In the beginning God created." Some modern translations believe that a more accurate translation is "When God began to create". This wording is the same as in a recent translation of the Torah, where the editors make this important point: "The implications of the new translation are clear. The Hebrew text tells us nothing about 'creation out of nothing', or about the beginning of time". Thus, there was no beginning of the sort modern science identifies as a *Big Bang*. The second part of the first sentence is traditionally translated as "the heavens and earth." There is no authority for using the word heaven, since the Hebrew word shamayim means sky. So an accurate translation would refer to sky and earth. The editors suggest the intended meaning is probably the entire universe.

The amount of scholarship devoted to gaining a better understanding of the Bible is enormous, and this brief survey indicates only some of the difficulties encountered in trying to determine the correct text. In spite of all this work, uncertainty in understanding some of the ancient words and statements remains. The three major ancient versions of the Bible—the Septuagint, the Masoretic text, and the Vulgate—differ in detail, and it is usually not possible to determine which text is to be preferred. This has led the biblical scholar Keith R. Crim (1994) to write:

> Because no original manuscripts of any of the books of the *Bible* are known to exist, and the ancient manuscripts that scholars use as the basis of translation differ widely at many points. . . . In spite of the wide agreement among scholars on textual questions, examination of the notes in the margin of modern translations reveals that scholars still differ as to which manuscript is the most accurate in a specific passage.

The opinions of biblical scholars are important in evaluating a fundamental tenet of creationists' belief, namely, the inerrancy of the *Bible*. Most likely no one in the last three thousand years has seen the original scrolls that were later combined as *Genesis*. So far as we know, those original manuscripts have joined the desert sands, and whatever truth they are thought to contain has to be accepted on faith alone; it is not confirmable on the basis of the available historical or scientific evidence. The only texts that creationists and all others have to rely on are the existing translations, whose history is checkered, to say the least. There are problems with the accuracy of copying, and there is uncertainty over the meaning of some of the ancient Hebrew words. Modern translations are the best that scholars have been able to provide, but the variations among them are appreciable. Still, since these texts are all that we have, if *creationists* and *evolutionists* want to compare their respective accounts of the diversity of life, it is to modern translations of Genesis that both groups must turn.

Two Creation Accounts in Genesis

Although the *Bible* is said to be the most widely read book in the Western world, few readers seem to notice that there are two very different accounts of creation in *Genesis*, one originating in the P scrolls and the other in the J scrolls. The P and J versions were combined to produce the first part of Genesis as we know it. Stephen Mitchell (1996) has recently provided a new, majestic, and beautifully written version of *Genesis* that identifies J, P, and other sources.

The story of how scholars came to the conclusion that Genesis has this complex composition is fascinating. It had been known for centuries that the *Bible* often presented two versions of the same event. This was a serious problem until the nineteenth century, when biblical scholars discovered that careful analysis of style can reveal significant differences between any two versions of a text. In the case of *Genesis*, this stylistic difference is most apparent in the original Hebrew, but subtle differences remain even in the translations available today.

The Hebrew word Elohim appears in the P version, while *Yahweh* appears in the J version. The translators of the King James *Bible* retained this difference by translating Elohim as "God" and Yahweh as "*Lord.*"

In the P version, the familiar account and the one most widely recognized by creationists, creation occurs over six days:

On the first day, when the earth was dark, wet, and formless, *God* (*Elohim*) created light and night and day.

On the second day God made a firmament (*heaven* or *sky*) to separate the waters above from the waters below.

On the third day, God separated land and water and created plants.

On the fourth day, God created the sun, moon, and stars.

On the fifth day, God created aquatic creatures and birds.

On the sixth day, God created other animals and man.

On the seventh day, God ceased all his work.

The much older J account of creation differs from the P account in that there is no reference to days. Some scholars interpret J to imply that all creation occurred instantaneously, not in six days as in the first account. A literal reading however, does imply the following sequence:

- The earth was barren and without plant life.
- Then the Lord *God* formed Adam from dust.
- The Garden of *Eden*, with all plants, was formed.
- The Lord *God*, noting that it was not good for man to be alone, formed wild animals and birds out of dust.
- None of the above being a satisfactory partner for *Adam*, the Lord *God* formed woman from one of *Adam's ribs*.

Note that in the J version the Earth already exists when the Lord *God* forms *Adam* from dust, and there is no mention of the creation of light, water, sky, land, or celestial bodies. By implication, then, these must already have been in existence. Note also that in the sequence of creation man is formed first, then the plants, animals, birds, and woman; in the P version aquatic creatures and birds are formed first, then all other creatures and man.

How are we to interpret these different accounts of creation? If we take every statement in the *Bible* literally, one or the other account must be correct, but both cannot be so. This is not a new theological problem. The early Fathers of the Christian Church tried valiantly to solve the dilemma long before it was realized that Genesis was woven together from two different sources. Most early theologians accepted what was later recognized as the P account as the creation story, but others found J preferable. Finally, it was agreed that both accounts must be accepted, since the Bible in its entirety was the Word of *God*. Saint Augustine and many others encouraged this point of view. Augustine's position was that nothing is to be accepted save on the authority of Scripture, since greater is that authority than all the powers of the human mind. It was hoped that an understanding of the apparently incompatible accounts of creation would come in time. Awaiting that day, it was necessary to believe both that God had created the earth and all else in six days and that creation had been instantaneous, as the second account was interpreted.

Andrew Dickson White, the famous historian, diplomat, and first president of Cornell University, gave a fascinating account of the early attempts to resolve the dilemma of the two accounts of creation (1898):

"Serious difficulties were found in reconciling these two views, which to the natural mind seem absolutely contradictory; but by ingenious manipulation of texts, by dexterous play upon phrases, and by the abundant use of metaphysics to dissolve away facts, a reconciliation was effected, and men came at least to believe that they believed in a creation of the universe instantaneously and at the same time extended through six days."

The presence in the *Bible* of conflicting creation accounts and other narratives is interpreted by biblical scholars as examples of compromises between different groups within the Jewish population—if you can't agree on which version to accept, retain the stories of both groups. Strict creationists reject this political analysis concerning the creation text. But if creationists insist on the inerrancy of the Bible, they cannot use the data of *Genesis* to support their position that P alone is the correct description of creation, since a second, conflicting account is also present. One cannot take *God* at His word in the P account and ignore what He says in J.

Noah's Flood

Most creationists believe that the great flood described in *Genesis*, commonly referred to as the Noachian flood, was a historic event that was worldwide in its extent and effect. Both the J and the P versions speak of a devastating flood that destroyed all life, including human beings, except for a few individuals of each species that were taken into the Ark by Noah and his family. In contrast with the accounts of creation in the first two chapters of Genesis, where P and J are kept separate, they are merged, which describe the flood with only a few inconsistencies. Nevertheless, biblical scholars are in general agreement that the separate versions can be recovered.

Both the J and P accounts of the flood begin with God's discouragement at what has happened to His creation, especially to human-kind, which has fallen into evil. *God* repents ever creating human beings (J) and decides to destroy them, along with most living creatures, in a terrible flood. *Noah* alone among humans meets with his approval. *God* instructs *Noah* to make an *Ark* so that he, his family, and a few mating pairs of all species can ride out the flood. The P account provides a brief description of the *Ark*, and it was huge: 300 cubits long, 50 cubits wide, and 30 cubits tall. A cubit is the distance from the elbow to the end of the longest finger. If a cubit is taken as 20 inches, the *Ark* would be about 500 feet long, 83 feet wide, and 50 feet tall. *Noah* was to take his extended family plus seven pairs of clean beasts, one pair of unclean beasts, and seven pairs of birds (J); or alternately one pair of all creatures (P). Everything else was to be destroyed. Food would be taken for the family and the beasts (P). Water, presumably, would not be a problem.

As soon as *Noah* and company were safely enclosed in the *Ark*, vast quantities of water poured upon the surface of the earth, from both above and below. It rained for 40 days and nights, according to J; according to P, the waters did not subside for 150 days and nights. The flood was so great that even the highest mountains were covered (J). Then the waters receded, and eventually dry land appeared. *Noah's* family opened the *Ark* door and along with all the creatures on the *Ark*, returned to the earth to be fruitful and multiply.

The creationists' belief that the flood described in *Genesis* actually occurred raises enormous problems, the major ones having been recognized for centuries. The *Ark*—a very big boat even by modern standards—was probably large enough to hold a few individuals from all the species of mammals and birds likely to have been known to the ancient inhabitants of the Near East, but it would have been totally inadequate for all of the many millions of species known to exist today. Supposedly all individuals not lucky enough to have obtained passage on the *Ark* were destroyed. The J account says that man, beast, creeping things, and birds are to be destroyed, whereas P says that all flesh is to be destroyed, and that all flesh in which there is the breath of life is to be destroyed. In both cases only the few individuals of each species that entered the *Ark* were to survive.

The most direct reading of these plans of *God* would seem to mean that all life not on the *Ark* was to be destroyed—even fishes, for they have the breath of life. Some scholars have argued that there was no need to worry about the fishes—they could handle a flood—but the huge number of insects and other invertebrates in existence seem to have been overlooked as possible passengers. And neither the J nor P authors considers plants. As far as we are told, neither plants nor seeds were taken on the *Ark*; yet when the flood subsided and Noah sent birds out to see if there was dry land, the dove came back with a freshly picked olive leaf (J). Although *God* promised that He would destroy every living substance that He had made, apparently not quite all living substances had been destroyed.

Yet another problem for which there is no natural solution concerns the creatures found in other parts of the world. After all, the entire world, including the highest mountain peaks, was apparently covered by the flood. That implies that all species from Australia, North America, South America, those parts of Europe and Asia distant from the Near East, sub-Saharan Africa, and the many islands of the world's oceans would have to travel many thousands of miles over land and sea to reach the *Ark*. And they had to do it in a week—the amount of time God gave Noah to prepare for the flood.

And finally, where would all that water, which would cover the highest mountains, come from, and where could it go when the rains stopped? Geologists have no idea.

Other Accounts of Origins

Every distinct tribe anthropologists have studied and every early civilization archaeologists have discovered seem to have developed a complex mythology to explain human life and nature. These represent attempts by pre-scientific and pre-literate societies to explain observations and experiences that were beyond the realm of their understanding. How could one explain the *sun*, for example? Where did it come from? Why did it appear and disappear with such extraordinary regularity? And of course where did human beings come from? Since every individual has a birth, could it be that human beings as a whole had a birth, a *creation*?

Many cultures have a story of creation, in contrast to a belief that human beings and the world have always existed. These explanations of creation invoke supernatural causes for events imagined to have occurred long ago. These explanations become part of the tribe's sacred traditions and are transmitted orally by the elders to the younger generations. A few examples of the creation accounts of ancient societies, mainly of the Middle East, are given here, and some of them show relations to the Genesis accounts of creation and the flood.

Mesopotamian Accounts of Creation

Ancient Mesopotamia (now Iraq), where the earliest known civilizations began, was the Land of the Two Rivers, the rivers being the Tigris and Euphrates. Several different Mesopotamian creation myths survive, the best known of which is *Enuma Elish*, named for the opening words that mean "*when on high.*" This account may date from around 2000 b.c., but the oldest known copies are from at least a thousand years later. The nineteenth-century British explorers Austin Henry Layard and George Smith and their fellow worker Hormuzd Rassam discovered the first of these at Nineveh, in the ruins of the library of the Assyrian king Ashurbanipal (669–627 b.c.). Although in some ways he was a brutal king, he was also a notable scholar with a splendid library—in this respect, an Assyrian Thomas Jefferson. The library contained not books but records of various sorts inscribed on clay tablets. These were prepared by inscribing the wedge-shaped grooves of cuneiform script into the still-wet soft clay of freshly prepared tablets. These tablets are almost indestructible, especially if baked—which happened when the king's palaces and library were burned.

Though unreadable when the tablets were discovered, cuneiform writing was later deciphered by Henry Rawlinson and others. These scholars found that the king had assembled a valuable series of texts, including many copies of very old Mesopotamian legends. Seven of these clay tablets contain almost the complete text of Enuma Elish. The story begins before creation with the presence of the divine parents, Tiamat the mother and Aspu the father, and their one son, Mummu. As is so often the case in early myths, the gods and goddesses are also natural objects. Thus Tiamat was a saltwater ocean and Aspu a freshwater ocean. Mummu probably represented mist. These three waters made up the physical world. A period of begetting by the gods resulted in a sizable group of divinities; one of special note was *Ea* (also known as *Enki* and *Nudimmud*), Tiamat and Apsu's great-grandson.

Later some of the younger gods became rather boisterous and fun-loving, much to the annoyance of the original divine pair. Tiamat showed a considerable degree of motherly understanding, but Apsu was determined to destroy the lot. That decision was most disturbing, quite naturally, to the targeted gods. *Ea*, however, put a spell on Aspu and at length killed him and assumed his position. *Ea* and his wife, *Damkina*, bore *Marduk*, who was to become the greatest god in the Mesopotamian pantheon.

Tiamat remained inconsolable after the death of Aspu and together with some other disgruntled gods, Kingu among them, plotted revenge against *Ea*. When the plan for battle became known, *Ea* and the gods loyal to him grew terrified of *Tiamat*—who could also become a dragon—and her followers. Negotiations attempted by *Ea* came to naught, and finally he turned to *Marduk* for protection. *Marduk* agreed to help his father, with the proviso that he would become the chief god and have control over the whole universe. That request was granted, but tentatively, since the gods were not entirely sure of Marduk's powers. To test his powers, they placed a garment in their midst and asked Marduk to make it vanish. He did. As a double check they also asked him to restore it. *Marduk* could do that, too, so the gods were assured of the strength of their champion—and they still had their garment.

Both *Marduk* and his great-great-grandmother *Tiamat* prepared for their titanic struggle. *Marduk* armed himself with bow, arrow, and club. He filled his body with flames, caught the four winds in a net, and made seven other winds. He mounted a storm-chariot, was preceded by lightning, caused a flood, and went searching for *Tiamat*. Her fellow gods were terrified of *Marduk*, but *Tiamat* was not. She accepted his challenge for a duel. *Marduk* caught *Tiamat* in the net containing the winds and caused them to blow into her body, which expanded her greatly. Then he shot an arrow through her mouth and down into her heart. That did it. Later he split *Tiamat* in two, the upper half of her body forming the sky and the lower half the earth. He also made the celestial bodies. All of the rebel gods were captured and forced to work in support of the victors. Tiamat's main supporter, *Kingu*, was taken before *Ea* and his arteries were cut. Kingu's blood was then used to create humankind. Human beings were forced to assume responsibility for serving the gods food and other necessities. This was a major task, since by this time the Babylonian pantheon had grown geometrically.

Biblical scholars find much of interest in *Enuma Elish*. For example, both *Genesis* and *Enuma Elish* accept that there was a creation—the universe as a whole was not eternal. Creation in both accounts starts with a watery chaos, a detail found in other Middle Eastern myths as well, followed by separation of the water into sky and land. Light forms before the sun and other luminaries are created in both accounts. There are, of course, major differences, such as the presence of only one *God* in *Genesis* and many in *Enuma Elish*. The parallels and differences have led scholars to pose four hypotheses: *Genesis* may have been derived in part from *Enuma Elish*; *Enuma Elish* may have been derived in part from *Genesis*; both may stem from a single, more ancient source; or the two may have no relation, their parallels being coincidental. There is no firm evidence so far to indicate which of these alternatives is correct, although many scholars prefer the hypothesis that some of *Genesis* derives from *Enuma Elish*.

Another Babylonian myth—the Epic of Gilgamesh, which may date to the third millennium b.c.—is remarkably similar to the *Genesis* account of the flood. *Gilgamesh* was a mighty hero—a human, not a god. There is even the possibility that the final redaction of Genesis might have been done in the library of that Assyrian patron of ancient legends, King *Ashurbanipal*. The discovery of the first clay tablets of the epic was made by Layard and his associates. One short section describes a flood that probably took place in Sumeria, the southern part of Mesopotamia. That area is a flat plain not much above sea level at the north end of the Persian Gulf. Moderate floods were common, and severe ones were known, especially the flooding of the Tigris River.

According to the Epic of Gilgamesh, long after human beings had been created they became a noisy rabble. The warrior god *Enlil* became so irritated that he decided to destroy the lot—every human being. However, the god *Ea* spoke to one man, the son of *Ubara-Tutu*, who lived in the city of *Shurrupak* and told him to build a large boat. It was completed in seven days and was rectangular, measuring 120 cubits on each side. Gold, wild and tame beasts, seeds, and supplies needed for the duration of the flood were brought on board, along with the family and the necessary craftsmen. For six days and nights it rained. By the seventh day all was silent; human beings not on the boat had been turned to clay. The boat came to rest on Mount Nisir. A dove was let loose, but not finding a place to rest, she returned to the boat. A swallow was then let loose, but it too returned. Later a raven was released. She found food and did not return. The doors of the boat were then thrown open, and the living creatures disembarked. The gods came, and a sacrifice was made. All were pleased, except for *Enlil*, who had planned for all human beings to be destroyed. However, *Ea* was able to placate him.

The resemblances between the floods in the Epic of Gilgamesh and in *Genesis* are striking, but as with *Enuma Elish* and the *Genesis* account of creation, scholars do not completely agree about the relationship between the two. Many similar stories, both oral and written, circulated in the ancient Middle East, and the priests of different cultures had opportunities to become familiar with many of them. This was especially so after 586 b.c., when the Babylonians captured Jerusalem and took many of the leading Jewish citizens back to Babylon, where they would have learned of the Mesopotamian myths.

Creation Myths of Egypt

The archaeological information available today indicates that civilization began in *Egypt* almost as early as it did in Mesopotamia—and almost as abruptly. The First Dynasty began in Egypt about 3100 b.c., after a long predynastic period. Hieroglyphic writing had been in use for a few hundred years. Thanks to the dry climate, which tends to preserve artifacts and even papyrus scrolls, a great deal of information is available to us about Egyptian history. Compared with Mesopotamia, where empires rose and fell, the history of Egypt was relatively serene. The center of Egyptian civilization, the Nile Valley, was protected to a considerable degree by deserts on the east, west, and south and by the Mediterranean on the north, all of which would have presented an obstacle to large-scale invasion.

The pantheon of *Egyptian* gods was immense and changed over the centuries of Egypt's ancient history. The total number of gods, great and small, has been estimated at about 3,000. The Nile Valley consisted almost entirely of agricultural villages. In addition there were few cities that were centers of religion and government. Each village had its special god that looked out for the interests of the inhabitants. The larger political divisions, the nomes, also had their own guardian deities. The two largest divisions, Upper and Lower Egypt, each had a special god with a distinctive crown. During those periods when Upper and Lower Egypt were united, a single god was shown with a composite crown. The name of the supreme god for the entire country varied, depending on the era: *Atum*, *Ra* (or *Re*), *Path*, or *Khepri*.

The surviving manuscripts and inscriptions deal mainly with the role of gods and goddesses in contemporary life. These roles were usually quite diverse and could vary with time and place. The deities were almost always associated with some animal or physical object. For example, a major god *Horus* was depicted as a falcon. *Atum*, a mighty god who created not only himself but the world and its inhabitants as well, was also the sun god, and depictions often show him with a solar disk on his head. *Osiris* was at one time the moon god but later was associated with life in the after-world; so, rather appropriately, he is often represented as a bearded mummy. *Hathor* had various roles, such as the goddess of the four quarters of the horizon. She was generally portrayed as having a human body and the head of a wild cow.

According to Egyptian accounts of creation that have come down to us from ancient times, at the very beginning all was water—just like in the creation accounts of *Genesis* and *Enuma Elish*. The goddess *Nun* ruled supreme, and she created the god *Atum*. In early dynastic times, Atum's mythic role changed and he became the original god, possibly self-created. While still alone, he mixed semen with dust and created his brother *Shu*, god of air and life, and his sister *Tefnut*, goddess of moisture. Another version has *Atum* creating his brother and sister from his spit. *Shu* and *Tefnut* produced *Geb* and *Nut*. When *Geb* and *Nut* were together, Nut on top and Geb on the bottom, father *Shu* separated them as sky and earth. They gave birth to the next generation, which included *Seth* and *Nephthys* as well as *Osiris* and *Isis* and their offspring, *Horus*. Continued reproduction by the gods eventually produced all human beings.

Egyptian gods endured for more than three thousand years—a considerably longer period than has elapsed from their eclipse to the present day. Egypt's political independence ended in 30 b.c., when it became a province of Rome. The ancient Egyptian gods slowly became less relevant as gods of other cultures took their place.

Creation Myths of Classical Greece

In the fifth century b.c. Greece's prominence was rising as Egypt's was declining. While many Greek philosophers were trying to understand the world in natural terms, most ordinary Greeks probably continued to accept the myths about their origins that had been handed down from ancient times—the sacred traditions of the tribe. One of the earliest legends is credited to the *Pelasgians*, whom the historian *Herodotus* believed to be the most ancient inhabitants of Greece. According to the Pelasgians, a goddess named *Eurynome* was the first deity. In the beginning there was nothing but a watery chaos. *Eurynome* divided the sky from the waters and danced on the waves. She captured the north wind and created from it a snake, *Ophion*. *Eurynome* and *Ophion* mated, and *Eurynome* changed into a dove and laid an egg. The egg was incubated by *Ophion* and when it hatched, there emerged from it animals, plants, earth, and the celestial bodies.

The mythology more commonly associated with the later Greeks passed from the oral to the written tradition with *Homer* and *Hesiod*, probably in the eighth century b.c. The creation myth in this tradition begins with the goddess *Gaia*, who emerged from chaos and produced the earth and oceans. With no biological helper she gave birth to *Uranus*, the god of heaven. Similar virgin births are not uncommon in early mythology.

Like *Nut* and *Geb*, *Gaia* and her son-husband then started producing gods. Their first offspring were the *Titans*, a varied lot that included most importantly the earth goddess, *Rhea*, and the evil god *Kronos*. Relations between the *Titans* and their father, *Uranus*, were far from good. *Gaia* tended to side with her children, as a good mother naturally would. *Kronos* decided to attack *Uranus* and did so while *Gaia* and

Uranus were making love. At that critical and exposed moment *Kronos* castrated Uranus. The divine penis and testicles were thrown into the ocean, and from them *Aphrodite*, the goddess of love, was born and came floating to shore on a seashell.

With *Kronos* as chief god, the earlier era of dominance by goddesses was over. He raped his sister, *Rhea*, and in time there were six children-gods. Many were eaten by *Kronos*, but later they escaped relatively unharmed from his body. *Rhea* was able to hide and thus save the last child, *Zeus*, who eventually overthrew *Kronos* and established himself as the ruling god on Mount *Olympus*. By this time a very large number of gods were endowed with all the good and evil qualities of human beings. They could be a rowdy lot, given to human passions, but with supernatural powers not available to ordinary mortals. The gods created various groups of human beings, but most were found wanting and were either hidden in the earth or sent to the Blessed Isles. The final creation was the Iron race, which was the progenitor of all humanity.

Andaman Islands Creation Myths

The last example of creation myths will be those of the former inhabitants of the Andaman Islands. These islands form an isolated 200- mile-long archipelago in the Bay of Bengal between India and Myanmar (formerly Burma). The native inhabitants, when first studied by anthropologists around the beginning of the twentieth century, had one of the least technologically developed cultures known, yet they had a complex mythology, so different from those of the Middle East. They used fire but apparently did not know how to kindle it.

Fish were one of their most important foods, but they did not have fish hooks; instead they used bows and arrows to shoot fish in the shallow coastal waters surrounding their islands. They had little contact with other people before the late nineteenth century, by which time they had developed not only a violent hostility toward outsiders but also a taste for them. The outsiders were mainly shipwrecked sailors, who when caught were promptly slaughtered and eaten. In the middle of the nineteenth century, in an effort to stop such killing, the British occupied the Andaman Islands, exposed the natives to civilization, and by the early twentieth century had driven them to extinction.

The British anthropologist Alfred Reginald Radcliffe-Brown made a classic study of the Andaman Islanders in the years 1906–1908. He found that they had many very different legends about the origins of human-kind, though there was no consensus among them. Here is one example:

The first man was *Jutpu* [meaning "alone"]. He was born inside the joint of a big bamboo, just like a bird in an egg. The bamboo split and he came out. He was a little child. When it rained he made a small hut for himself and lived in it. He made little bows and arrows. As he grew bigger he made bigger huts, and bigger bows and arrows. One day he found a lump of quartz and scarified himself. *Jutpu* was lonely, living all by himself. He took some clay [kot] from the nest of the white ants and molded it into the shape of a woman. She became alive and became his wife. She was called *Kot*. They lived together at *Teraut-buliu*. Afterwards Jutpu made other people out of clay. These were the ancestors. *Jutpu* taught them how to make canoes and bows and arrows, and how to hunt and fish. His wife taught the women how to make baskets and nets and belts, and how to use clay for making patterns on the body.

One of the variations on this origin legend accounted for the origin of animals: "*Sir Prawn* once got angry and threw fire at the people [the ancestors]. They all turned into birds and fishes. The birds flew into the jungle. The fishes jumped into the sea".

The world of the Andaman Islanders was enormously complex. Most things and processes of nature had not only a physical being but a spirit as well. Whereas the belief systems of the major religions invested supernatural powers in one or more supernatural beings, the Andaman Islanders believed that natural objects also had supernatural powers. The legends that recorded their history involved a constant interplay of the forces of nature and of human beings. There was no elaborate pantheon of gods of the sort that characterized the religions of Mesopotamia, Egypt, or Greece. Some spirits might be more powerful or more important than others, but that was all.

After Radcliffe-Brown left the islands, he developed a working hypothesis to explain what he had observed: "Customs that seem at first sight either meaningless or ridiculous have been shown to fulfill most important functions in the social economy, and similarly the tales that might seem merely the products of a somewhat childish fancy are very far indeed from being merely fanciful and are the means by which the Andamanese express and systemize their fundamental notions of life and nature and the sentiments attaching to those notions".

Myths and the Human Heart

In societies of the developed world—now characterized by electronic technology, an exploding understanding of nature, a wealth of literature in all fields, and the desire to base conclusions on rational data—the term myth has a pejorative ring. It has not always been so. In the long span of human history before literature and literacy were widespread, the myths of the tribe provided the only taste of what is still essential in our lives today: knowledge apart from the chase and the hearth. *Myths* explained the forces of nature as well as the complex interrelations of human beings with one another and with the rest of their living and nonliving environment.

In spite of their tremendous variety, myths share some interesting features. Most obvious is an element of the supernatural: myths invoke gods, demons, devils, or angels with powers not available to mere humankind. The gods are not constrained by the laws of nature. Their supernatural powers are in fact violations of what seems natural to us and what we are able to do. Yet if those powers are used for our welfare, we feel less lonely and less personally responsible for what is the fickleness of fate.

Creation myths are almost universal among the world's cultures. This is astonishing. It would be far easier, and perhaps more logical, to suppose that the world has always been in existence. But even the modern science of cosmology, which studies the origin and structure of the universe, proposes a creation story: that a Big Bang started the universe around 15 billion years ago. It is intriguing to ask why the notion of a creation even arose. We may never know the answer, but possibly the observable fact that some plants grow from tiny seeds and some animals develop from tiny eggs may have given our ancestors the idea that the world as a whole has an origin.

One commonality of creation myths is the movement from simplicity to complexity. In the words of David and Margaret Lemming (1994): "The basic creation story, then, is that of the process by which chaos becomes cosmos, nothing becomes something. In a real sense this is the only story we have to tell".

So what does the present survey of Genesis and of other creation stories reveal about the hypothesis of supernatural creation as described in *Genesis*? I believe it puts us in a position to examine this claim with the same rigor we would apply to any document from any culture. The conclusions drawn are as follows:

1. There are two conflicting accounts of creation in Genesis, according to the J and P texts. One or the other may be correct, but both cannot be. There is no logical mechanism for making a choice, and there is no scientific evidence in support of either version.
2. In ancient times, and to some extent today, different groups of people have developed unique stories of creation by a supernatural force in the remote past. There is no logical way of choosing one "correct" creation story from among the many accounts available.
3. Stories of creation are part of the sacred traditions of individual groups, and each group accepts its own traditions on faith. Those accepting the P or J version of creation in *Genesis*, or any other story of creation, do so on the basis of belief, not on scientific evidence.

Theologian and biblical scholar Alan Richardson (1953) addresses the evolution versus creation conflict this way:

When we understand the nature of the Genesis parables, we shall no longer suppose that there can be a conflict between "*science*" and Genesis. . . . Genesis is not a scientific account of how the world came into existence; if I want to learn how this happened, I must go not to Genesis but to science. It is misleading even to speak of Genesis as "*pre-scientific*," for *Genesis* is not concerned with scientific questions at all; it is not a collection of the guesses of primitive men at answers to scientific questions. It is dealing with matters beyond the scope of science; its theme is man's awareness of his existence in the presence of God, his dependence upon and responsibility toward *God*. This high theme is dealt with in the only satisfactory way in which the human mind can deal with it, that is, in religious symbols. We shall miss the whole point of Genesis if we either take the parables of Creation, the Fall, and the rest, literally or look upon them as primitive guesses at scientific or philosophical truth.

Most scientists would find Richardson's statement acceptable because it distinguishes between what science is and what it is not. Science can say nothing about a god, gods, or any other supernatural phenomenon; the tools and methods of science have been honed to explore the natural world, not the supernatural realm even if there is such a thing. By the same token, Genesis brings no scientific evidence to bear on the history of life over time, the age of the Earth, or the diversity of life that we see today. It provides no acceptable explanations for the remarkable biochemical similarities in the cells of microorganisms, plants, and animals, for the universality of the genetic code, for the peculiarities of the geographic distribution of species, for the basic structure of major groups of organisms and the variations shown in the individual species in the group, or for the cycles of life that maintain a rough biological and chemical equilibrium in the environment.

The *Bible* has basic things to say about morality, spirituality, and religious practice, and for many people its wisdom provides structure and meaning to their personal lives. It is not, however, a textbook of science, and no serious scholar or theologian pretends that it is. There is a saying attributed to Galileo, and possibly Saint Augustine before him, that goes to the heart of the matter: "The Bible tells us how to go to *Heaven*, not how the heavens go."

15

COPYING AND CREATING

In comparing the development of organisms to a creative process, such as painting a picture, it may appear a very important distinction: creativity involves originality and inventiveness that seem without parallel in biological development. An artist does not continually paint the same picture again and again: each creation is different from the previous one. As Leonardo said, 'The greatest defect in a painter is to repeat the same attitudes and the same expressions'. Organisms, though, seem to develop with much greater consistency. To be sure, every individual is slightly different from the next; even identical twins do not look exactly the same. But the extent of variation seems much less than that between different original paintings. The development of a mouse or an oak tree appears to be more highly circumscribed and defined from the outset than the process of creating a picture.

Perhaps the development of organisms is more like copying the same picture again and again rather than a creative process. After all, we use the word reproduction in both art and science with this type of comparison in mind. In art, it refers to the process of making copies from an original; in biology it is the production of new individuals every generation. The outcome of both processes is comparable: you end up with lots of things that look quite similar to each other—many copies of the Mona Lisa or many rabbits.

The extent to which this comparison between reproduction in art and biology is valid. We shall see that there is an element of similarity between the two types of reproduction, but there is also a fundamental difference that will eventually bring us back to the issue of how development compares with creative processes.

Reproduction in Art

To *reproduce* a work of art, you need to be able to copy it in some way. There are various ways of doing this. Look at a class of children being taught how to copy a leaf, taken from a teachers' handbook on drawing of 1903. Every child is copying the leaf on the blackboard with remarkable consistency, conjuring up our worst images of Victorian discipline. Although this example might be on the extreme side, it shows how the notion of copying is generally associated with discipline and slavish imitation rather than imagination. Making a good copy of a picture seems to require excellent technique and rigour but not the creativity that might go into producing an original.

There are other ways of reproducing a picture, apart from copying by hand. Most modern reproductions of paintings are made by photographing the original. Here copying has become almost entirely a technical exercise: a question of mastering the camera and printing process. The picture is automatically transformed into a negative image, from which as many positive prints can be manufactured as needed, so long as you have the appropriate equipment and expertise. An equivalent type of copying is used to reproduce sculptures, by first taking a mould, a negative, and then making a cast to get back to something that looks like the original sculpture. In all these cases of art reproduction, whether it is done by hand alone or with the aid of various devices, there is always a process of copying from an original or template. The final goal is already there before you start—the leaf on the blackboard is there for all to see. The aim is simply to make something that resembles it as closely as possible. In the most successful case, you end up with a replica that might almost be substituted for the original. In creating, however, the aim is not to produce a replica of what is already there. You might of course be inspired by a beautiful woman sitting before you, but the final picture is not simply a copy of the woman; it is a two-dimensional image on a canvas that the artist has created for the first time. We would have no difficulty in distinguishing the painting of *Mona Lisa* from the person in the flesh. The aim of the artist is not to produce something that could ideally replace the subject of the painting, but to create a special type of image. Copying and creating are different sorts of process. Which of these does biological reproduction come closer to?

Reproduction by Division

A good place to begin looking at biological reproduction is with microbes. Any glass of pond water or clump of soil is teeming with microscopic organisms, many of which consist of single cells. These unicellular organisms are of the order of one-thousandth to one-hundredth of a millimeter long. Although tiny, they are far from simple. They are able to grow by taking up and metabolizing various nutrients from their environment and can even exhibit what might be called '*behaviour*', such as swimming towards or away from a particular chemical. These organisms come in many shapes and sizes and are remarkably diverse in their range of life-styles, occupying habitats from hot springs to the deepest reaches of the oceans. Nevertheless, they are all constructed on similar principles. Each has a membrane going all the way round it that keeps the fluid contents of the cell from spewing out. The membrane provides a boundary, allowing the cell to maintain its individuality, separating it from the outside. This means that anything that enters or leaves the cell has to pass through the membrane. The cells are far from being just bags full of simple fluid, though. They contain a range of complex structures and molecules—some of which we will encounter later on—that are essential for their various activities.

The activity of these unicellular organisms that most concerns us here is their ability to reproduce by a process called cell division. When an individual cell grows to a certain size, it starts to narrow in the middle. The narrowing continues until eventually the cell becomes divided into two separate cells, like a drop of water splitting in two. Under ideal conditions, a single-celled bacterium, such as *Escherichia coli*, which lives in your gut (normally without harmful effects), can grow and divide once every twenty minutes, allowing it to multiply to more than one million individuals in a mere seven hours. The problem of reproduction for many unicellular organisms therefore boils down to the question of how a cell can grow and divide into two.

Reproduction for unicellular organisms is not quite as monotonous as this: most of them also undergo some sort of sex once in a while. This usually involves two individual cells coming together—in some cases completely fusing—to produce a hybrid cell. It can be compared to cell division in reverse, two cells becoming united as one, rather than one cell dividing in two. Why unicellular organisms, or any organisms for that matter, have sex is a complicated question that has been argued about extensively in scientific literature. We need not go into it here, except to say that one undoubted consequence of sex is that it allows an exchange of genetic information between individuals. This means that even for unicellular microbes, the process of reproduction involves more than just cell division. (In some unicellular organisms, such as *Acetabularia*, reproduction is further complicated by their undergoing significant changes in shape during their life cycle.)

Multicellular Organisms

Fascinating though the reproduction of microbes might be, it seems to be a far cry from the way you or I reproduce. We cannot go forth and multiply by splitting ourselves in half. Our style of reproduction is more complex than that of a unicellular microbe, but nevertheless there is a common element that links the two. Both microbes and ourselves are based on the same type of building blocks, cells, but whereas individual microbes may comprise a single cell, we are made of many billions of cells; we are multicellular organisms.

The realization that plants, animals and microbes are based on the same fundamental cellular units is one of the greatest unifying discoveries to have been made in biology. It has a curious history that went in fits and starts. We can begin back in 1665, when Robert Hooke described the microscopic structure of cork as a network of tiny chambers, or cellulae. The Latin word cella means 'a *small room*' and we still use it in this way when talking of a hermit's cell or prisoner's cell. The term seemed appropriate at the time because Hooke was not looking at living cells but at their relics: the cells in cork are dead, so he was seeing the skeletal outline of the previously living cells of a plant. The cell was therefore originally thought of as being a passive container rather than a living entity. This concept gradually shifted as the contents of cells were looked at more carefully, so that the cell eventually became identified as a living unit of life rather than just a vessel. A key advance made in the early nineteenth century was the discovery that most living cells contain a small body, called the nucleus, suspended in the fluid of the cell. By comparing the nuclei of plant and animal cells, the botanist Matthias J. Schleiden and animal physiologist Theodor Schwann, realized in the 1830s how both types of cell might be formed on very similar principles. Schwann recalls that at the time he had been working on the nerves of tadpoles and frogs and had noticed nuclei in the cells of a particular structure called the notochord: One day, Mr Schleiden pointed out the important role that the nucleus plays in the development of plant cells. I at once recalled having seen a similar organ in the cells of the notochord, and in the same instant I grasped the extreme importance that my discovery would have if I succeeded in showing that this nucleus plays the same role in the cells of the notochord as does the nucleus of plants in the development of plant cells.

Plant and animal cells contained a similar looking nucleus, suggesting to Schleiden and Schwann that both types of cell might have been generated by a common process. This led them to propose that both plants and animals were constructed from the same elementary units, cells, formed by a single universal mechanism. In other words, trees, frogs, worms and people were all made of cells that had arisen in the same way. Unfortunately, although they did a great service in unifying plant and animal biology, the mechanism that Schleiden and Schwann believed to be responsible for cell formation turned out to be wrong. They thought that cells formed anew by growing within pre-existing cells, and their strength of conviction (particularly Schleiden's, for whom modesty was not a strong point) misled biologists for more than twenty years. It eventually became clear that all cells multiply by division rather than forming anew: every cell in a plant or animal has arisen by division of previous cells.

In parallel with the work on multicellular plants and animals, the cellular nature of microscopic organisms was also becoming dear in the mid-nineteenth century. These tiny creatures, each comprising a single cell, shared many features with the individual cells of multicellular organisms. The cell therefore became the fundamental unit of all life.

This unity can now be viewed as reflecting a common evolutionary past. The *first cells*—the common ancestors of all life on earth—are thought to have arisen as unicellular organisms about three and a half billion years ago. For the next three billion years or so, life continued to be dominated by unicellular creatures. The distinction between plants and animals is thought to have occurred whilst life was still in t is unicellular phase. Unicellular plants sustained themselves using energy trapped from sunlight, through a process called photosynthesis. Unicellular animals survived by feeding off others. These different life-styles led to specializations in cell construction that are still evident in the cells of plants and animals today. The self-sufficient life-style of plants was compatible with having a hard protective casing or cell wall. Animal cells, however, had to retain mobility and flexibility to catch and engulf their prey and could not afford to be surrounded by a cumbersome rigid wall. When complex multicellular plants and animals evolved, about half a billion years ago, these basic differences in cell construction were retained. Both types of cell have an outer membrane surrounding the cell fluid, called *cytoplasm*, and a nucleus within. The nucleus is surrounded by its own membrane, containing pores that

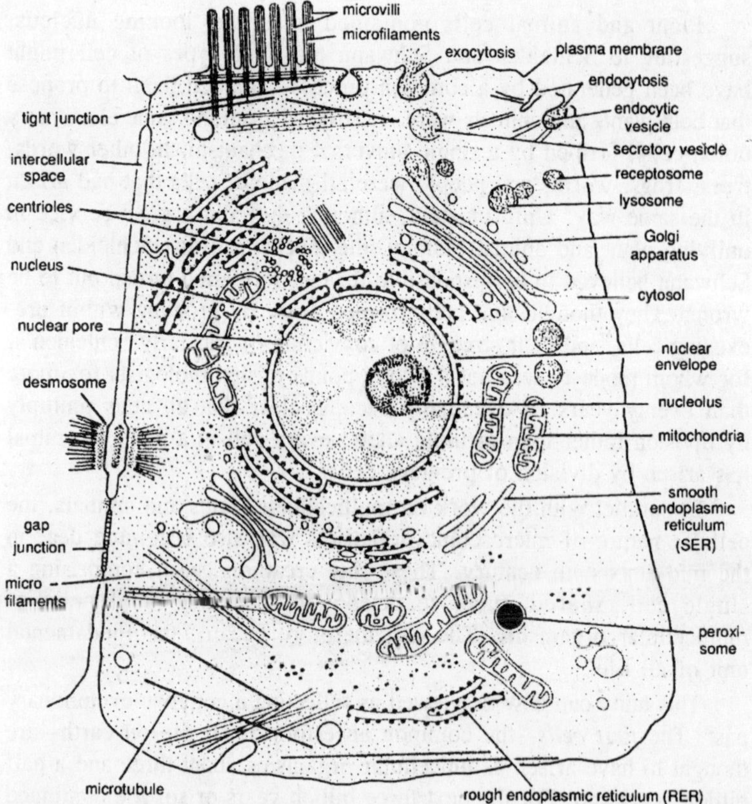

Fig. 15.1. Ultrastructure of a typical animal cell.

allow molecules to pass between the *nucleus* and *cytoplasm*. In addition, plant cells produce a hard outer coating, the cell wall (cell walls provide the major ingredient of paper and wood). It was these walls—the protective covering—that Robert Hooke saw when he described cells for the first time by looking at cork down a microscope.

The differences between the cells of plants and animals have had many repercussions on the way these organisms are constructed. Multicellular plants are supported by a meshwork or lattice of cell walls that extends throughout their body. They can continue to grow and develop extra parts throughout their life, like ambitious neighbours who are forever adding extensions to their house, because each new addition carries its own internal lattice of supporting cell walls. In contrast, animals are supported by a framework of bones or a toughened outer covering, made by specialized cells in restricted locations of the

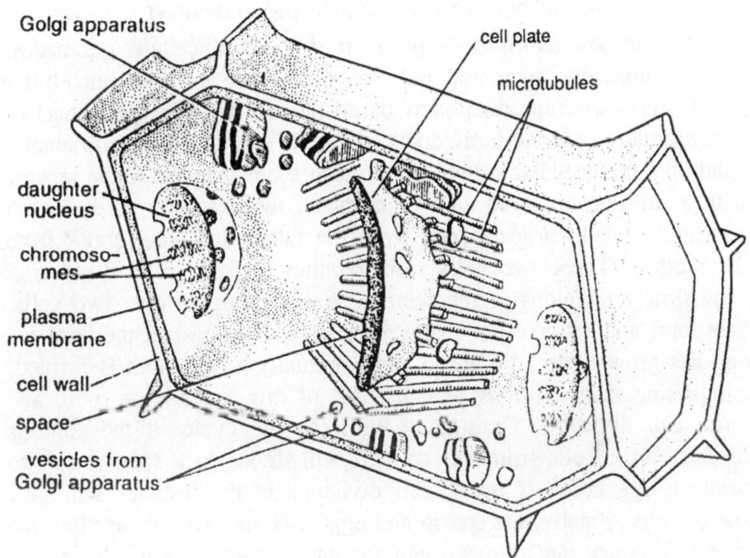

Fig. 15.2. Schematic representation of a cell of a higher plant.

body. Development tends to be concentrated in the early phase of an animal's life when all its parts are formed in the appropriate arrangement. A child is born with all the essential limbs and organs in place and this basic arrangement is maintained for the rest of its life. (Some animals do undergo a major reconstruction during their life by going through a second phase of development. For example, when caterpillars turn into butterflies they undergo redevelopment within the confines of a pupa by a process called metamorphosis.) Another consequence of these differences in life-style and construction is the distinct way that plants and animals respond to their environment. An individual plant explores its environment through growth: extending, branching, spreading and invading the space around it whilst rooted to the same spot. An animal achieves similar results through moving its whole body. A moth flies towards light, whereas a plant grows and twists in its direction. Animals can protect themselves by running, hiding or fighting. A plant is far less able to avoid damage but can tolerate enormous losses to its body because it has the ability to keep growing, to the point that it can become a chore to mow the lawn every week. Some animals, such as crabs, are able to regenerate lost parts, like a missing limb. However, in these cases, the limb is regenerated at its original position, whereas plants generally respond by growing extra parts rather than replacing on a new-for-old basis.

Reproduction Through Development

We can now return to the problem of how multicellular organisms, such as humans, mice and oak trees, reproduce. Like unicellular organisms, their reproduction is based on cell division but instead of making many separate individuals, the cells stay together to gradually build up a multicellular individual. We can summarize the whole process with a life cycle. In the case of humans, for example, parents each contribute a cell: a *sperm cell* from the father and an *egg cell* from the mother. These two cells fuse together to form a *fertilized egg*. This slowly develops in the womb, first dividing to give two cells, then four, and so on. After many more rounds of division, the fertilized egg has grown into an embryo and eventually a new adult is formed, comprising many billions of cells. All of this depends on more and more cell divisions. Continuing with the life cycle: if the adult is female, some cells from her ovaries will divide in a special way to produce egg cells. If male, cell divisions in the testicles will give sperm cells. Finally, the sperm and egg cells are brought together and fuse to produce the fertilized egg for the next generation, closing the cycle. The life cycle involves alternation between a single-cell phase, the fertilized egg, and a complex multicellular adult phase. The two are linked by numerous cell divisions and an occasional sexual fusion.

A similar cycle underlies the reproduction of a flowering plant. When a grain of pollen from one flower lands on the female part of another flower, a sperm cell from the pollen fuses with an egg cell in the mother. The fertilized egg then divides repeatedly, doubling the number of cells each time until a tiny plant embryo forms. The process temporarily stops at this point and the plant releases the embryo with a protective hard outer coat, in the form of a seed. In the right conditions, as when the seed is planted in the ground, cell divisions resume in the embryo so that it eventually grows into a new plant, with flowers which produce more egg cells and pollen grains. As with humans, single-cell phases alternate with multicellular phases. One difference between flowering plants and humans is that many flowers are hermaphrodites, producing both male and female organs; this sometimes allows an individual to fertilize itself if pollen lands on female organs from the same plant. (There are also some plants, like willows and stinging nettles, that are more like us in having separate sexes.) Some multicellular organisms, such as aphids and dandelions, can side-step the requirement for sexual fusion during the life cycle and are able to develop from unfertilized eggs, although in these species there is still alternation of single-cell and multicellular phases.

To understand the mechanism by which multicellular organisms reproduce, it is not enough to know how a cell can grow and divide; we also need to know how a single cell, the fertilized egg, can give rise to a multitude of different types of cells in the complex arrangements that form the mature individual. The human body contains many organs and tissues, each made of various types of cells— nerve cells, blood cells, hair cells, etc.—each arranged in a precise way. These different cell types can be distinguished by characteristics such as size, shape, structure and behaviour. By behaviour some of the more dynamic properties of cells like whether they move, grow, divide or even die. In a similar way, the various parts of a plant are made up of many different cell types with different properties, although unlike those of animals, plant cells do not usually move relative to each other because they are fixed in position by their cell walls.

The problem of development is to understand how the complex pattern and arrangement of different cell types that make up a mature organism can arise from a single cell in a consistent way each generation. This problem applies to multicellular organisms, and not to unicellular organisms that reproduce by simple cell division. Throughout this book I shall use the term development in this sense: to refer to the process whereby a single cell gives rise to a complex multicellular organism.

Proteins as Guiding Shapes

The properties of every plant or animal cell depend on the types of *protein* it contains. We normally come across proteins as part of our diet. Proteins, though, play a much more pervasive role in our lives than this might lead us to believe. All the processes in the body, such as digestion, secretion, moving, sensing and thinking, depend on the activity of different types of protein molecule. Without proteins we would not be able to do anything. The most important feature of protein molecules that allows them to encourage all these things to happen is their shape. The way in which a proteins shape can influence events is rather central to this book.

If you put an empty bucket outside and let it fill with rain, you might say the bucket is holding the water. Obviously this does not mean it is actively doing anything about the water, trying desperately to keep it all together. The shape of the bucket leads to the water being held in a particular way as long as the rain pours down to fill it. The bucket facilitates or guides the way that the downpour of water is collected. Without the bucket being there, the water would never heap up on its own to form a bucket-shaped mound. The process is driven by an energy source that is outside the bucket: the rain pouring down from above, or more remotely, the sun's energy that evaporated water from the earth's surface and led to the formation of clouds. We could imagine more complex combinations of shapes, such as a mountain-side covered with buckets, perhaps connected together by a network of other shapes, in the form of tubes or pipes that guide rain water from bucket to bucket and finally into a reservoir. Further pipe shapes could guide the water to drive a turbine and generate energy in a different form. Each shape facilitates one course of events rather than another but does not itself provide the required energy to drive the process along.

In a similar way, each cell contains many thousands of different types of proteins, each one with a different shape, according to the process it guides. These processes are at a sub-microscopic scale, the scale of molecular reactions. Molecules in a fluid are always on the move, continually jostling around very rapidly and bumping into each other. The higher the temperature, the faster molecules move around; and at the temperature needed to sustain life, there is quite a commotion in the cell's interior. For a molecular reaction to happen, say for molecules A and B to join together, the molecules need to come together in the right way. Normally, when A and B happen to bump into each other, nothing might happen because they do not meet in a suitable manner and they just career off again into the distance. But suppose A encounters a large molecule, a protein, that has a shape with a nice little pocket that A fits into very comfortably. We could imagine, for example, that the pocket in the protein matches the shape of the A molecule, like a lock matching a key. The A molecule may stick to the protein and not career off. If the protein has another nearby pocket that matches molecule B, then when B is bumped into, it will also tend to stick. There is a reasonable chance that the protein will have both A and B stuck to it at the same time and, if they are

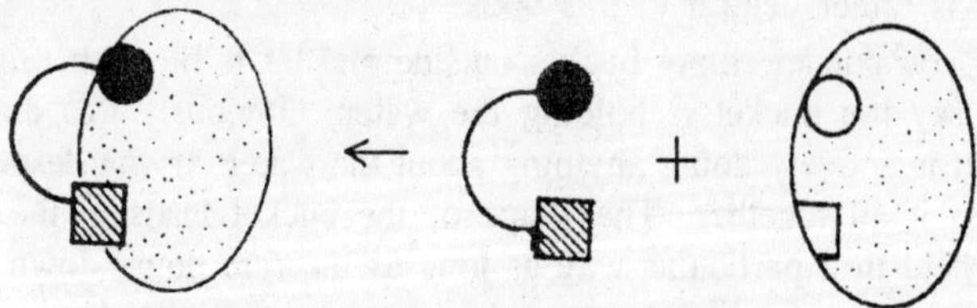

Fig. 15.3. Combination of enzyme with the substrate in form ES complex according to Fisher's hypothesis.

held in the right way, they will react with each other, joining up to form a new molecule, C. In this way, the shape of the protein, the structure of its pockets and crevices, can facilitate a reaction: A and B coming together to make C. Once C has formed, it may leave the protein, freeing up the pockets to join another pair of A and B molecules. Because the protein is not consumed by the reaction but simply helps to guide it along, it is said to act as a *catalyst*. All of the molecular events catalyzed by a protein happen extremely quickly: a protein may promote 1000 reactions like this every second. This is because the molecules move around and react with each other at such a mind-boggling rate.

Proteins that catalyze these sorts of reactions are called *enzymes*. In the example shown, C is the product of the reaction; but it is also possible for the reaction to go in the reverse direction, breaking C down into A and B. The direction which any reaction takes will largely depend on the energy involved in making or breaking chemical bonds between the molecules and on the amount (concentration) of A, B and C molecules in the cell.

Matching the shape of a molecule by a protein is not just a question of complementing its three-dimensional shape. Molecules also have small electrostatic charges distributed over them, so that some regions of a molecule tend to be more positively charged and others more negatively charged. In matching the shape of molecules, proteins also match their distribution of charges, so that a positive region of a molecule lies next to a negative region of the protein. These charges can be very important in slightly deforming the shape of molecules, by pulling or repelling parts of them, and thus facilitating particular types of reaction. To simplify matters, the three-dimensional structure of a molecule and its particular charge distributions.

Proteins do not provide any energy to drive reactions; they are just catalysts by virtue of their particular shapes. Through its compatibility with various molecules, the shape of a protein can encourage or facilitate one reaction occurring rather than another. Each different type of reaction usually needs a protein with a different shape. The protein that helps A and B come together would not help D and E; they would need their own protein to help them along. Every cell therefore needs many thousands of different types of protein, each with a distinct shape, to guide its numerous internal reactions. Proteins with various shapes will be mentioned throughout this book, so it is very important to remember that the role of a protein, the process it guides, is an automatic consequence of its shape. It is not actively 'doing' anything other than facilitating one thing or another happening, just as a bucket helps water collect in one place rather than another. Nevertheless, a protein does this or that as a form of shorthand. It is so much easier to say a protein 'does X' than 'its shape facilitates X happening'. This should be taken in the same spirit as saying a bucket 'holds water' rather than 'its shape facilitates water assuming a bucket-like conformation:

As with the bucket filling with water, the energy that drives all these proteins ultimately comes from the sun (with the exception of some bacteria that obtain their energy from inorganic compounds). Solar energy is captured within the cells of plants, where it is guided to produce sugars from carbon dioxide and water. The light energy is effectively converted into a different form of energy, stored in the chemical bonds of the sugar molecules. The chemical energy and components in the sugars are then channelled in all sorts of different directions, by other types of protein in the plant, to make molecules such as carbohydrates, fats and more proteins. These products are in turn essential for sustaining animal life: when an animal eats a plant, the energy and chemical components of the plant are channelled by the animal's proteins to make its own carbohydrates, fats and proteins. In other words, the energy that drives the internal reactions of animals comes from the food they eat, which ultimately depends on energy from the sun.

There is one more key feature of proteins: their shape is not completely rigid but can change, depending on which other molecules happen to be bound to them. When a molecule binds in a pocket, it can cause a change in the protein's overall conformation or shape. In most cases, when the molecule leaves, the protein will return to its original shape. These reversible changes in protein shape underlie much of our behaviour. The movement of every muscle in your body depends on countess muscle proteins changing their shape back and forth very quickly. Similarly, the transmission of electrical signals in your brain and nerve cells depends on rapid reversible changes in the shape of particular proteins. As with the other processes I have mentioned, the energy to drive all these events does not come from the proteins themselves, but ultimately comes from the sun.

Given that the combination of protein shapes in a cell is responsible for many of its properties, a major part of trying to understand development has to do with explaining how some cells of the body come to contain different proteins from others. Why is it that cells forming in the brain region have proteins appropriate to brain cells whereas those in the liver region have proteins relevant to liver cells? There are thousands of cell types in the body, all arranged in a very precise manner, so the problem of how each comes to have its own particular spectrum of proteins becomes rather daunting.

Making a Biological Copy

To understand how the structure of proteins is determined, another key molecule, DNA is introduced. *DNA* is a very long molecule made up of two strands; it is double-stranded, the way that string is often made from two intertwining strands. Each strand is itself made of a sequence of molecular subunits, called bases, of which there are four different types, symbolized with the letters A, C, G and T (the letters actually stand for the chemical names of the bases, *adenine*, *cytosine*, *guanine* and *thymine*). The bases are strung together in a particular order, just as letters are ordered in every word; although unlike our

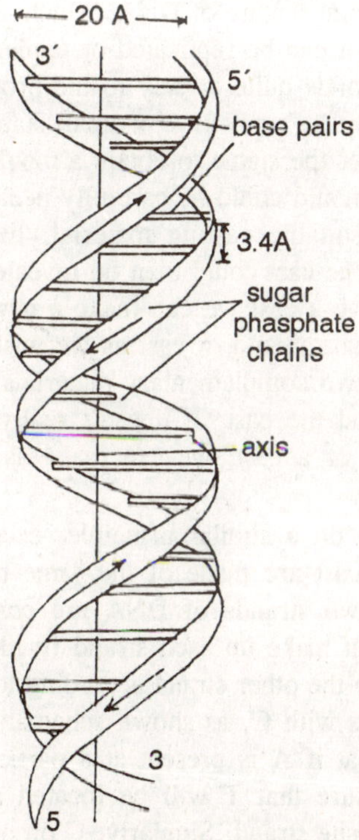

Fig. 15.4. Structure of DNA.

alphabet, which has 26 letters, there are only 4 types of letters in DNA. Nevertheless, a DNA molecule can carry a lot of information with its four-letter alphabet because it is enormously long.

Most cells have several DNA molecules of various lengths, each individual DNA molecule being called a *chromosome*. In many organisms the chromosomes are in pairs, with members of the same pair having the same length and a comparable DNA sequence. The numbers and lengths of chromosomes are often different between species. For example, humans have 23 pairs of chromosomes in the nucleus of each cell, giving a total of 46. Fruit flies have 4 pairs of chromosomes per cell, and *Antirrhinum* (snapdragon) plants have 8 pairs. A typical human chromosome has about one hundred million bases in it, making it several centimeters long. If we were to magnify this to the width of string, with the bases spaced at 1 mm intervals, it would stretch for about a hundred kilometers. Although very long, each DNA molecule is very tightly packed, allowing all 46 chromosomes to be stored within the nucleus of a single cell.

The most important feature of DNA, which gives it such a unique place in life, is that it can be replicated or copied. This happens in a manner that corresponds quite closely to the process of reproduction in art. One way to make replicas of a small statue would be to pour a rubber mixture over the statue to obtain a mould. After the mixture has set, the rubber mould could be carefully peeled off the statue and then filled with a suitable casting material, like plaster, that will eventually harden. The cast could then be revealed by peeling off the mould. Once we have a cast we can make a new mould from it and so continue to replicate mould or cast as we wish. All that is needed to make replicas is two complementary materials that can match each other: the mould and the cast. It doesn't really matter whether we start from a mould or a cast, we can make as many copies as we want.

DNA replicates on a similar principle, except that in this case the mould and the cast are made of the same material. Like a cast and a mould, the two strands of DNA are *complementary* to each other. The bases that make up each strand fit like jigsaw pieces into their counterparts on the other strand according to the following rules: A fits with T, G fits with C, as shown schematically in the enlarged part. This means that if A is present at a particular position on one strand we can be sure that T will be located at the corresponding position on the opposite strand. Similarly, C on one strand will always be opposite G on the other. Knowing the sequence of bases on one strand, we can therefore predict the complementary sequence of the other strand.

To replicate DNA, the two strands need to be separated, just as we need to take the mould off the cast if we want to make new copies of a statue. Once separated, each strand can then be used as a mould or cast to produce a *new complementary strand*. The separate strands act as templates, ensuring that only matching bases are lined up along their length to form a new matching strand. We eventually end up with two identical double-stranded DNA molecules, having started off with only one, so there is now an additional pair of casts and moulds. This mechanism of replication follows quite naturally from the structure of DNA, the complementarity between its two strands. This is why the discovery of the structure of DNA, by James Watson and Francis Crick, was such a watershed: it pointed the way to how *biological copying* might occur. As Watson and Crick wrote in a famous passage towards the end of their paper in 1953: It has not escaped our notice that the specific pairing we have postulated immediately suggests a possible copying mechanism for the genetic material:

The replication of DNA does not occur spontaneously: it requires an input of energy and some guidance (just as statues don't proliferate without someone separating and pouring the casts and moulds). There are specific proteins that guide the energy and components needed for

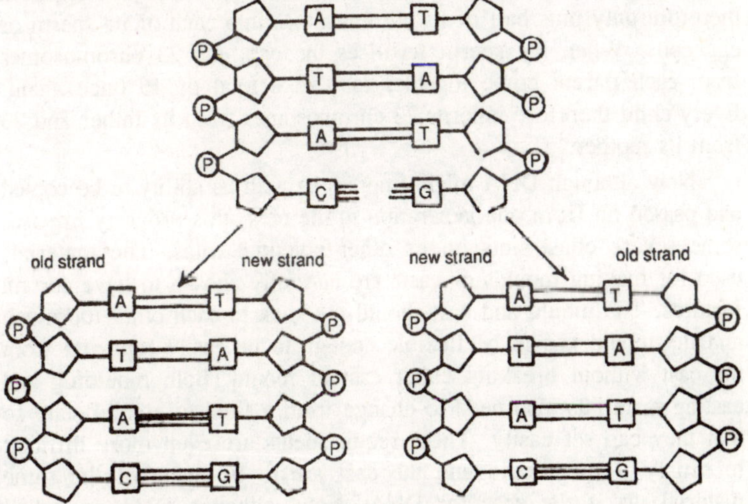

Fig. 15.5. Process of DNA replication.

DNA copying, ensuring that the individual bases that make up the new DNA strands are incorporated efficiently and effectively (the bases are themselves made by chemical reactions catalyzed by yet other proteins). It is important to emphasize, though, that although these proteins help to ensure that DNA is copied, they do not themselves provide the information that is stored in the DNA sequence. The person who pours the moulds and casts when copying a statue does not provide the information in the statue, its shape and form. He or she is merely allowing an existing structure to be copied. In the same way, the proteins that catalyze DNA copying do not determine the sequence of the DNA—that is already provided—but simply help the sequence to get replicated.

The process of DNA replication normally occurs every time a cell divides so that each daughter cell inherits one set of DNA molecules. A dividing human cell containing 23 pairs of chromosomes would therefore contribute a replica of all 46 chromosomes to each cellular offspring. This sort of division, mitosis, is typical of most cells in your body. There is a less common type of division, meiosis, which is nevertheless very important because it is essential for sexual reproduction. In the case of humans, meiosis occurs in some of the cells of the testicles and ovaries. During the division of these cells, the two members of each chromosome pair become separated from each other, so that you end up with progeny cells, sperm and eggs, having only 23 rather than 46 chromosomes in their nuclei. Each parent therefore only puts half of its chromosomes into each of its sperm or egg cells. When the sperm fertilizes the egg, the 23 chromosomes from each parent come together to give a total of 46 once again. Every child therefore inherits 23 chromosomes from its father and 23 from its mother.

Now although DNA might impress us with its ability to be copied and passed on from one generation to the next, this property imposes some severe constraints on its other potential roles. The materials used for making moulds or casts are carefully chosen to have certain features. The mould and cast should not stick to each other too much and the mould has to be flexible enough to be easily removed from the cast without breaking either cast or mould. Both moulding and casting materials also need to change from a fluid to a solid state so that they can set easily. These requirements are even more difficult to satisfy if the mould and the cast are to be made of the same material, as is the case for DNA. Now, although DNA meets all these requirements very well, it is at the expense of other properties, such as being able to fold up into complicated shapes. The features of DNA which make it so suitable for being copied render it much less useful for doing anything else in the cell. This means that for the information contained in DNA to have any significance for the behaviour and properties of cells, it needs to be converted into a different form: the currency of proteins.

From DNA to RNA to Proteins

Proteins are made of molecular subunits, called *amino acids*, strung end to end. There are 20 different types of amino acids, 20 letters in the protein alphabet. The shape of the protein, and hence what it guides, depends on the particular sequence or order of its amino acids. A typical protein might contain several hundred amino acids strung together in a very specific order. Although they are joined together in a linear chain, the amino acids nudge, pull and push against each other, owing to their own individual shapes, adjusting their positions so that the chain folds on itself until the protein ends up with a stable arrangement or overall shape. Every type of protein has its own diagnostic sequence of amino acids and hence its own shape that allows it to catalyze particular events in the cell.

Proteins and DNA are similar to each other in both being built from sequences of molecular subunits—amino acids and bases respectively—but are very different in other ways. Whereas proteins are able to fold up into all sorts of wonderful shapes, allowing them to promote particular events in the cell, DNA is structurally much more limited, being more like a long piece of string that sits in the nucleus. On the other hand, this wearisome feature of DNA is precisely what allows it to do something that proteins cannot: DNA can get copied very easily. The information in DNA—the sequence of its bases—an get replicated and passed on every time a cell divides. By contrast, once a protein is made it cannot be used as a template to make more copies of itself. Considered alone, proteins and DNA each suffer from a major deficiency, but by working together they can make up for each other's weaknesses.

We can start by notionally dividing the sequence of a long DNA molecule or chromosome into shorter stretches, say a few thousand bases long, each stretch being called a gene. A chromosome can be 100 million bases long, so we could theoretically divide it up into many thousands of genes, following each other along its length. Each gene can be thought of as a word, a few thousand letters long, written in the four-letter alphabet A, T, G and C. A chromosome of 100 million bases will therefore contain many thousands of such gene words along it. It has been estimated that humans have about 70000 genes in each set of 23 chromosomes, while fruit flies have about 12000 genes in their set of 4 chromosomes, and a plant such as *Arabidopsis* (a small weed commonly found in gardens) has about 25000 genes in its set of 5 chromosomes.

Like words, genes are characterized by the particular sequence of letters they contain. The most important part of a gene, as far as proteins are concerned, is a stretch called the coding region. This region contains information, coded in the four-letter alphabet of DNA, that when properly translated can lead to a particular type of protein being made. There is a precise correspondence between the sequence of bases in the coding region and the sequence of amino acids in the resulting protein. Thus, each type of protein is encoded by a distinct gene. Once you know the nature of this correspondence, called the genetic code, you can convert any sequence of DNA into a protein sequence, just as once you know Morse code you can convert a series of dots and dashes into a word written in letters. The genetic code is organized in triplets, so that a set of three DNA bases, such as ATG, corresponds to a particular amino acid, called methionine in this case. Another triplet, GCA, corresponds to a different amino acid, called alanine. In this way, the DNA sequence in the coding region is translated as a series of consecutive triplets, to specify a sequence of amino acids in a protein. This means that a stretch of DNA has to have three times as many subunits along its length as the protein it codes for: a coding region 300 bases long will be needed to give a protein of 100 amino acids. There are also some triplets with roles similar to punctuation, indicating where translation of the DNA information should begin or end.

We need not be concerned here with all the details of how the information from a gene is translated into a protein, but there is one important further aspect that needs to be covered. The translation of information from DNA to protein does not happen directly but involves an intermediary: another type of molecule called RNA. Like DNA, the RNA molecule is made of a string of four types of bases, each type of RNA base matching the shape of one of the DNA bases. But whereas DNA is double stranded and very long, an RNA molecule is made up of a single strand and tends to be much shorter: a typical RNA molecule is only a few thousand bases long, compared to the hundreds of millions of bases that can go into DNA.

For a gene to make a protein, the coding region of a gene is first copied or transcribed into RNA molecules. The process of transcribing DNA into RNA is not too different from the way DNA is replicated. The two strands of DNA are separated, and one of them acts as a template for the assembly of a string of complementary RNA bases in the appropriate order. For any gene, only part of the DNA—the sequence covering the coding region—is transcribed into RNA in this way. The RNA copies are then free to leave the nucleus and enter the surrounding cell fluid, the cytoplasm. Here, the information in the RNA copies is translated into a sequence of amino acids, to make a protein. Depending on its shape, the protein may remain in the cytoplasm, or it may enter the nucleus, or it may be secreted out of the cell. To make a protein, information therefore has to be transferred from DNA to RNA to protein.

There is an interdependence between DNA and proteins. As far as the structure of each type of molecule is concerned, information flows in the direction from DNA through RNA to protein. This means that if the DNA sequence of a coding region is altered, the amino acid sequence of the encoded protein may consequently change. The reverse does not apply: a change in the sequence of a protein has no direct effect on the sequence of DNA. Structural information does not flow from proteins back to DNA. On the other hand, the information in proteins—their shape—is needed to guide the various events in a cell, including the replication of DNA, *transcription* from DNA to RNA and translation of RNA into proteins, as well as numerous other processes that occur in a cell. This mutual dependence between DNA and proteins is based on the very different properties of each type of molecule: they are specialized in complementary ways.

You might be wondering how such a system ever evolved in the first place, because neither proteins nor DNA can function alone. One view is that this system evolved as a specialization from a more primitive form of life—existing billions of years ago—that was not based on DNA and proteins but on RNA. The properties of RNA are in many ways intermediate between DNA and proteins: like DNA it is able to be copied, but it is more versatile in its shape and can catalyze a limited number of chemical reactions in the cell. Perhaps, then, life was originally based on a single type of molecule, RNA, that was able both to replicate and to promote particular reactions. According to this view, DNA and proteins evolved later on as a specialization that allowed these different aspects of life to be carried out more effectively.

Reproduction in Biology and Art

Let me summarize the story so far. In art, reproduction involves the process of copying. In exploring biological reproduction, we have also come across a copying process: the replication of DNA. There is an original template—a set of DNA molecules—that is dutifully copied and passed on from parents to offspring every generation.

Nevertheless, although we have revealed a copying process involved in biological reproduction, what is being copied is very remote from a complex organism. Forty-six balls of string, each marked with a sequence based on four colours, denoting A, G, C and T, are hardly the same as a human being. There is currently an international effort, the *Human Genome Project*, aimed at determining precisely the DNA sequence of all human chromosomes. The human sequence will not be uniquely defined, because all individual humans have a slightly different sequence, but it should be possible to get a sequence that is typical of human DNA. Even when we know this enormously long string of letters, however, it will not tell us how we are made. To understand our reproduction as complex organisms, we have to be able to relate this sequence in some way to the process by which a fertilized egg cell develops into a living human being.

The relationship of a DNA sequence to a fully developed organism is not captured by any comparison with artistic reproduction: humans are not related to DNA as copies are to an original. If development were like copying, we would have to imagine that there was some sort of miniature human in the fertilized egg that was being used as the original. We have seen, though, that the subject of biological copying is not a tiny human, but something with an entirely different structure: DNA. This raises the fundamental problem of how this sequence of DNA is related to the complex biological form that eventually develops from the fertilized egg.

Now although this has no parallel with reproduction in art, there is, an artistic activity that comes near to it in many ways: the process of creating. By creating, the highly interactive process that goes on when, say, an artist paints an original picture. Unlike copying, the final aim is not there in front of the artist to begin with. The picture emerges through an interaction between the artist, the canvas and the environment. The process by which a DNA sequence eventually becomes manifest in a complex organism has many more parallels with the interactive process of creating an original than with making a replica or copy. Of course, there is still the issue of open-endedness: the extent to which the development of organisms is more highly circumscribed than painting a picture.

From DNA to Organism

To unravel the problem of development, we will need to understand more about the *DNA language*: how the sequence of bases in the DNA molecule is related to the organism that eventually develops from a fertilized egg. Before proceeding any further, a general method for deciphering the meaning of DNA that will be employed throughout this book. Deciphering DNA is rather like learning a completely new type of language. As children, we learn language in several ways. One is by association—people pointing to objects and saying the relevant word like '*table*', '*chair*', '*flower*'. This may help to convey the significance of a noun but it does not cover other aspects of language such as verbs and sentence structure. We learn these more easily through listening to and exploring how words are used. By continually hearing and using strings of words in various contexts we eventually start to get a more comprehensive knowledge of the meaning and structure of language. To understand the meaning of DNA we have to somehow find a way to establish gene associations and uses. It turns out that there is a method for doing this but it is almost the reverse of the way we usually acquire language.

Imagine a society that is identical to ours except that there are no tables. This may be returned to as a mutant society, deviating from ours in one respect: the absence of tables in this case. No one would ever feel the need to refer to tables in this society, and a word for table would not be part of the language. If external observers were to compare this society with our normal one they would notice two things that distinguished it: (1) there are no tables and (2) the word table is missing from the language. By associating these two things the observers might conclude that the word table is used to refer to the objects that are present in the normal but missing in the mutant society. In this way the significance of the word table could be guessed at. The important point about this way of deciphering a language is that the tables and the word table go together. Remove one and the other automatically vanishes. It is a form of learning by association; an indirect form of pointing, like a '*spot the difference*' game where you show two pictures that only differ in one detail, such as the apples being missing from a tree. As long a you can associate the missings objects with a word, in this case apples, the significance of the word can be deduced.

In our mutant society, we first removed the tables and then found the word table was not necessary. But what if it also worked the other way round? Suppose that when you remove the word table from a language, the tables disappeared. In this case the mutant society is the direct result of removing the word. Obviously this is not the way our language works: we cannot remove words and watch objects disappear. But although such a notion might sound peculiar to us, from the external observers' point of view, the situation is the same as before. They are presented with the same two societies and would still come to the conclusion that table refers to the objects missing from the mutant society.

This sort of reverse logic is a very important method used by biologists to learn the significance of particular DNA words. The missing words correspond to *mutations* in genes. Although DNA is normally replicated with remarkable fidelity, occasionally a mistake is made in the copying process, such as a wrong base being put in (say an A instead of a G) or a small stretch of DNA failing to get copied and therefore being deleted in the daughter molecules. When such a mutation occurs it can have lasting consequences because it will be faithfully replicated and passed on to all subsequent DNA copies— just as if we were to scratch a cast and then make a mould from it, all subsequent casts made from that mould would carry the scratch. Mutations are normally very rare: typically one error is made for every billion bases copied. In some conditions, as when DNA is exposed to ultraviolet radiation, the rate of mutation can increase due to the DNA having been damaged. That is why excessive sunbathing can be harmful: it results in DNA mutations in the skin cells that can sometimes lead to cancers. In this case, the mutation is restricted to the person in which it occurs (i.e. it is not transmitted to offspring) because it affects only cells that remain in his or her body. If, however, a DNA mutation arises in cells of the germline, which gives rise to sperm or eggs, it can get passed on to the next generation. When a sperm or egg cell carrying a mutation contributes its share of DNA to a fertilized egg, the individual that develops from the fertilized egg will carry the mutation in all cells of its body. Should the individual have offspring, it will transmit the mutation through its own sperm or egg cells to the next generation (actually the mutation will be passed on to only half of the offspring because each sperm or egg cell produced by the individual will carry only a half-share of chromosomes). In this way, the mutation may eventually accumulate in a population of individuals.

Some mutations in the germline will have little effect on organisms that later develop because they change a relatively unimportant bit of DNA, but other mutations may have significant consequences. For example, if the mutation is in the coding region of a gene, it may change the sequence of amino acids in the encoded protein and hence its shape. The altered protein may no longer be able to function properly in guiding a particular process in the cell. With this process missing or defective, the organism that develops may also be altered. In this way, a DNA mutation can lead to an observable difference in the organism, and because DNA is replicated, the difference will be heritable. The inherited differences in eye or hair colour, flower or leaf colour, height, shape and the numerous other characteristics that distinguish individuals are a consequence of variation in their DNA sequence that arose at some point from mutations. Essentially all heritable variation, the raw material for evolution, is a consequence of mutations in DNA.

Mutations are very important for getting at the meaning of DNA because they provide us with a way to link DNA with an organism's appearance. By associating the alteration in an organisms characteristics with a mutation in a particular piece of DNA, we get an indication of what that piece normally signifies. It's like learning a language backwards, removing words and seeing which objects disappear. When a lot of normal red-flowered *Antirrhinum* (snapdragon) plants breed and a mutant with white flowers is obtained, it can be told that a gene signifying red has been altered.

In all these cases, the effect of the *mutation* is the opposite of what the gene normally signifies because it reflects a defect in the gene. Mutations typically show what happens when the action of a gene is negated or removed rather than displaying its positive effects. This can cause some confusion because mutations and genes are usually named according to how they were first noticed rather than what they normally signify. For example, when a mutant fruit fly was noticed with white instead of the normal red eyes, it was said to have a mutation in the white gene, even though normally this gene is involved in making the eyes red. So in many cases, a gene's name indicates the opposite of what it normally signifies for the organism because of the topsy-turvy way that we discover its meaning.

Although the study of mutations in this way can be very revealing, it by no means gives us a complete picture of the DNA language. If all we did was point to objects and say their names, a child might learn nouns by association but very little else. The way words are used and combined in various ways to form meaningful sentences would not be appreciated. There are also further complications with our method of learning by association. In a mutant society without tables, there would be knock-on effects on other objects; for example there would be fewer chairs and no table-cloths. In trying to work out the meaning of table by association, we might get confused and think the word also refers to some missing chairs and table-cloths. If we only have simple associations to go by, it will sometimes be difficult to distinguish these knock-on effects from the primary defect in our mutant. We therefore need an additional method for working out the meaning of DNA: we need to determine how the genes are used during the life of the organism.

We have already taken some steps in this direction by considering how genes code for proteins, via an RNA intermediate. To get further, we need to follow the activity of those genes that are particularly relevant to an organism's development. One of the main problems with trying to do this is in identifying the small stretch of DNA, the gene, that relates to a particular feature of development. In the 1970s and 80s, methods were developed for *gene cloning*, which allowed specific genes to be isolated from organisms and multiplied in bacteria. Once isolated in this way, *specific genes* together with their corresponding RNA and protein molecules could start to be studied in great detail. This eventually led to a much better understanding of how genes are used during development. One of the main purposes of this book is to explain how this approach has given us much deeper insights into what the *DNA language* means.

In practice, the study of genes in this way has been carried out on relatively few organisms. The choice of which organism to study has often been a source of lively debate and rivalry among biologists. Ask a biologist why he or she has chosen a particular organism for study and you will hear of its many advantages. The arguments biologists use to justify their choice of organism have varied through history, as biological problems and techniques have changed. A favourite organism for early developmental studies was the chick. As Ernst Haeckel pointed out in 1874:

Hens' eggs are easily to be had in any quantity, and the development of the chick may be followed step by step in artificial incubation. The development of the mammal is much more difficult to follow, because here the embryo is not detached and enclosed in a large egg, but the tiny ovum remains in the womb until the growth is completed.

Although the chick was a favourite for a long time, other organisms, such as frogs, newts and sea urchins, became very popular in the late nineteenth and early twentieth centuries. This is because scientists became interested in seeing what happened when bits of embryos were surgically removed or interfered with. Frogs and sea urchins are particularly amenable for this sort of approach because their embryos are readily accessible and can survive such assaults rather well. More recently, the detailed study of genes and mutations imposed yet new criteria on the choice of which organism to investigate: organisms that can grow and breed in very large numbers and have a short generation time became desirable. The fruit fly, *Drosophila melanogaster*, often to be seen circling around dustbins or wine glasses, is a particular favourite because it is easy to maintain hundreds of flies in a bottle, and the time between generations is only two to three weeks. Important work on the fruit fly was started in about 1910 by Thomas Hunt Morgan but it was not until the 1980s, with the advent of gene cloning, that it became the subject of extensive molecular studies. Other animal species that have been intensively studied in this way include the mouse, nematode worm (*Caenorhabditis elegans*) and zebrafish.

Many plant species also offer advantages for genetic studies because they are easy to self-fertilize, cross and breed in large numbers. It was considerations like these that led Gregor Mendel to use peas for his initial studies on heredity in 1865. More recently, many plant biologists have chosen to work on a garden weed, *Arabidopsis thaliana*. Working on a weed is not as surprising as it might first sound because some of the desirable properties of an organism for genetic studies, such as rapid growth and short time between generations, are also those that make a good weed. In the late 1980s it became apparent that these advantages could be combined with molecular approaches, and research on *Arabidopsis* spread through plant laboratories like gold fever. However, for one particular plant, which also happens to be the *Antirrhinum majus* (snapdragon), because it turns out to have several advantages for the study of genes involved in flower development.

To summarize, the reproduction of multicellular life involves the development of a complex organism from a fertilized egg. At the heart of this process are DNA molecules that get copied every time a cell divides, much as a work of art might be reproduced. But this raises the fundamental problem of how DNA, a linear sequence of bases, is related to the elaborate three-dimensional organism that finally develops from the fertilized egg. To address this, we need various methods for deciphering the DNA language, revealing its meaning and significance for the organism. In practice, these methods have been applied to a few well chosen plants and animals. It is some of the basic lessons that have been learned from these studies, and what they say about the relationship between development and other processes, such as human creativity.

INDEX